THE
HEART
—OF THE—
INTERNET

Also by This Author

The Network Revolution. Berkeley: Penguin, 1982; London: Penguin, 1983.

Computer Message Systems. New York: McGraw-Hill, 1984.

Electronic Meetings (with Robert Johansen and Kathleen Spangler). New York: Addison-Wesley, 1979.

Dimensions. Chicago: Contemporary, 1988.

Confrontations. New York: Ballantine, 1990.

Revelations. New York: Ballantine, 1991.

Forbidden Science. Berkeley: North Atlantic, 1992.

FastWalker (a novel). Berkeley: Frog Ltd., 1996.

Les Enjeux du Millénaire (in French). Paris: Hachette, 1998.

The Four Elements of Financial Alchemy. Berkeley: TenSpeed, 2000.

THE
HEART
—OF THE—
INTERNET

An Insider's View of the Origin and
Promise of the On-Line Revolution

JACQUES VALLEE, PH. D.

HAMPTON ROADS
PUBLISHING COMPANY, INC.

Cover design by Marjoram Production
Cover art © 2003 Colin Anderson/Brand X Pictures/Picturequest

Hampton Roads Publishing Company, Inc.
1125 Stoney Ridge Road
Charlottesville, VA 22902

434-296-2772
fax: 434-296-5096
e-mail: hrpc@hrpub.com
www.hrpub.com

If you are unable to order this book from your local
bookseller, you may order directly from the publisher.
Call 1-800-766-8009, toll-free.

Library of Congress Cataloging-in-Publication Data

Vallee, Jacques.
    The heart of the Internet : an insider's view of the origin and
promise of the on-line revolution / Jacques Vallee.
        p. cm.
Includes bibliographical references and index.
    ISBN 1-57174-369-3 (6 x 9, Hardcover w/Dust jacket)
    1. Internet--Access control. 2. Information society. 3. Information
technology--Social aspects. 4. Electronic commerce. I. Title.
    TK5105.875.I57V355 2003
    004.67'8--dc21
                                                        2003004568

10 9 8 7 6 5 4 3 2 1
Printed on acid-free paper in the United States

Acknowledgments

In the course of my career, I have been privileged to work with several of the seminal thinkers and builders of the Internet, men like Doug Engelbart, Larry Roberts, and Paul Baran. Any book about the network is homage to their creativity and perseverance. I also have been guided by numerous computer entrepreneurs and by venture capitalists who financed their projects: Dan Lynch of Interop fame, Stephens Millard at Telebit, Metricom and Com21, Fred Adler at Euro-America Ventures, Brian Pinkerton at Excite, Peter Banks at the University of Michigan and XR Ventures, and my son Olivier have contributed to this book in many ways. Their advice was indispensable.

Several close associates have urged me to document my own experiences with the early days of networking, a subject that gets increasingly blurred as journalistic reconstructions and marketing hype keep rewriting its history. Indeed a reviewer of Internet books on Amazon.com offered this apt remark:

> Considering that the history of the Internet is perhaps better documented internally than any other technological construct, it is remarkable how shadowy its origins have been to most people, including die-hard Net-denizens!

I am grateful to my friend and colleague Graham Burnette and to Russell Targ, Robert Chartrand, Peter Beren, Paul Saffo, Robert Johansen, Rich Miller, and especially to Frank DeMarco and Richard Leviton at Hampton Roads Publishing for providing input, clarity, and encouragement. Vint Cerf provided a wealth of detail and helped me correct historical and technical inaccuracies throughout my early manuscript. Connie McLindon and Bob Kahn were generous with their time when I interviewed them in Washington.

Most of all, my wife Janine, herself a computer professional who has lived these challenges and adventures by my side, has been my constant source of inspiration and support.

Contents

Prologue ix
Part One: The First Explorers
1. Getting Out of Our Sphere 3
2. The Digital Society: Solid State or Grapevine? 26
3. Arpanet Genesis 43

Part Two: Making the Planet a Better Place
4. The Birth of a New Culture 73
5. Growing the Grapevines 90
6. An Unfinished Revolution 112

Part Three: The Betrayal of the Internet
7. The End of Innocence 125
8. Building the Future Mesh 139
9. They Want Well-Trained Humans 151

Part Four: How We Can Save the Dream
10. Four Essential Principles 165
11. Your Personal Countermeasures 177
12. To Create a System 187
Endnotes 192
Index of Individuals Cited 195
Index of Topics 197
About the Author 201

Illustrations

Chart 1: The Early History of Computers 15

Chart 2: History of the Arpanet 75

Chart 3: History of the Internet 103

Chart 4: History of the Web 116

Fig. 1: The Z3 Computer in 1941 16

Fig. 2: Alan Turing 17

Fig. 3: The Mark I Computer in 1949 20

Fig. 4: Programming the ENIAC 22

Fig. 5: The SRI Console in 1968 50

Fig. 6: Early Version of Engelbart's System 51

Fig. 7: Paul Baran 53

Fig. 8: The Author at the SRI-ARC Computer Lab, 1971 62

Fig. 9: The Author with Doug Engelbart and Elizabeth "Jake" Feinler, 2000 .63

Fig.10: Four Pioneers of Computer Networking 77

Fig.11: Members of the Forum Team, IFTF, 1972 85

Fig.12: The InfoMedia Startup Team in 1980 86

Prologue

A popular government without popular information, or the means of acquiring it, is but a prologue to a farce or tragedy or perhaps both.

—James Madison

Important developments in human history may be motivated by greed or fleeting visions of glory, but those that succeed have a vast spiritual dimension that defines them forever in the mind of the world.

Aviation, before it became a multibillion-dollar business, was the dream of private inventors like the Wright Brothers and pilots like Lindbergh. We fly on Boeing 747s, but our imagination relates best to the *Spirit of St. Louis* and to the poetry of Saint Exupéry's *Little Prince*.

We are not simply pawns in a technological world or helpless actors in an economic game. Long-term human actions are dictated by spiritual imperatives. We are inspired by the heroism of pioneers and by visions that take us beyond our human condition.

So it is with the Internet.

Most Americans discovered networking about 1995 and were immediately inspired by its seemingly magical ability to link people all over the world, irrespective of space and time. In a few short years, the technology has transformed business practices and personal lives with mythical power. But in this unprecedented expansion, something has gotten lost: the dreams and visions of the people who built the network. With that loss comes the dangerous temptation to abandon its potential benefits or to allow obsolete business interests to reduce it to another method of commercial manipulation of human beings as helpless consumers.

The story of the evolution of computer technology as an adventure of the human spirit has never been told, from the first elementary structures that emerged from the Second World War to the all-embracing web of today.

With over 40 million hosts on the Internet, 300 million websites, and 72,000 newsgroups, it is remarkable that no book is available on the complete history of the amazing network that is changing economic, social, and spiritual patterns all over the world. There are some 600 million online users around the planet today. Yet the only volumes published about the origin, design, and impact of the technology are either journalistic reconstructions (often based on accurate interviews, but hardly touching on the deeper issues) or difficult technical texts that give us no inkling of what will come next.

Even the information published by Internet companies about their own history is often wrong. Witness a hilariously misleading contest on AOL's home page on August 8, 2002: "Who invented the Internet?" was the topic of the quiz. And the multiple-choice answers were (1) the FBI, (2) Al Gore, (3) AOL, and (4) Tim Berners-Lee.

Given these alternatives, the closest answer to the truth would be "Al Gore," because he did play a role in authorizing some of the early Internet budgets. Tim Berners-Lee, the inventor of the web, was barely in high school when the Internet was conceived!

Because of this ignorance, and because the spirit of the Internet has never been clearly articulated and pursued, we are in danger of experiencing the dark side of the technology before we can reap the rewards of its full development.

I was finishing an early draft of this book when airplanes hijacked by terrorists destroyed the twin towers of the World Trade Center and caused death and devastation at the Pentagon. It rapidly came to light that the Internet had been used in the planning of the attack. On October 3, 2001, a French judge ordered the arrest of twenty-seven-year-old Kamel Daoudi, accused of managing e-mail communications for the Djamel Beghal network that was preparing to attack the U.S. embassy in Paris. The group was sending along innocent-looking images containing hidden messages, a method known as steganography. According to French counter-terrorism agents, the order to strike was expected to come from Kandahar in Afghanistan through text embedded in photographs. Governments around the world quickly reacted to such threats with measures to monitor or restrict communications through the web.

Even before the tragic events of September 11, 2001, the question of the Internet's future and its impact on our lives was hotly debated in terms of privacy invasion, civil liberties, the legitimate needs of law enforcement, and national security issues.

The attack on America has escalated the problem, demonstrating the central nature of the Internet not only for business and science but for national strategy and the survival of human spirit.

The issue of society's response to terrorism is beyond the scope of this book, but we can no longer assume that old definitions of privacy and security—the definitions that were adapted to traditional mail and telephone communications—apply to the world in which we now live.

Mechanisms for controlling illegal uses of the Internet have failed. Does that mean that our individual rights, guaranteed by a Constitution that was written in precomputer days, should be restricted? And how vulnerable is our economy as it increasingly relies on the Internet for commerce, business, and personal communications, and for culture?

These are the questions we urgently need to address. To reduce the issue to simple antagonism between "good" business users and "evil" hackers, or between government snooping and freewheeling expression (sharp G-men in dark suits versus gentle hippies with graying beards) represents a too-common pitfall. In times of crisis, such caricatures only make the problem worse. I invite you to go deeper, invoking more than a sterile quarrel about laws and standards.

In this book I want to introduce you to the true pioneers of the Internet, not in an abstract fashion but in the context of the history of computers to which they have added a brilliant new chapter. I have been fortunate in meeting most of them and in working closely as part of their teams. You will witness their expectations and their failures, their daily struggles against inertia and bureaucracy. More important, we will recapture their vision. They were not simply trying to achieve a technical marvel, but a better planet.

The founders of the Internet industry sought to free the human spirit from the boundaries of time and space. They wanted to give future generations access to new forms of community interaction and boundless opportunities for creativity and innovation. This is the vision we urgently need to recapture, before the vast network that has become known as "the web" turns into a tool for exploitation and control of the human mind, consumer patterns, and the behavior of citizens.

In the Internet of today, data-mining and customer profiling are fast replacing free information sharing as a fundamental mechanism. Obsolete structures of government, Hollywood, and major industries are reasserting their power over the new media in the name of law enforcement, efficiency, or market share. In the process, the spirit of the Internet is being betrayed. What most users perceive, rightly, as a wonderful opportunity for exploring a wider world is increasingly dominated and subtly controlled by the same handful of corporations that dominate our news media, commercial television, and publishing. The inspiration for universal access to other minds is being forced into narrow commercial channels.

Ironically, this trend threatens the full development of commerce on the web just as it hampers the blossoming of the network as a sophisticated infrastructure for human spirituality and free expression. Unfortunately, users are unaware of the tools that are being put into place to shape their behavior in this new world. Nor do they realize that they, too, have enormous powers: the network offers ways of preserving open access to the extraordinary variety of ideas, artistic forms, and opportunities for interaction that the technology has made possible.

This book will uncover these mechanisms for you. And I propose that we reexamine the very process that has led from early notions of the role of computers all the way to the Internet revolution, to ask how we can, and should, influence its future direction. In the process, we will meet the architects of this new world of information to learn from their experience.

The visit begins in Silicon Valley.

Menlo Park, December 2000

The International Center at SRI International is a glass and marble building that stands proudly over the trees of Ravenswood Avenue in quiet Menlo Park, California. Over the last five decades the Stanford Research Institute (SRI) has served as one of the country's most distinguished think tanks, solving scientific and technical problems for the aircraft industry, banks, computer firms, and the federal government. The sharp devices that stick out of the wings of airliners were invented by SRI in the 1950s to dissipate static electricity; the dollar amount you inscribe on your personal checks is deciphered by software it developed; it invented the Hydrocushion braking system for the railroad cars of the Southern Pacific; its experts labored on xerography, satellite imagery, pattern recognition, robotics, and many aspects of life sciences at a time when the term "Silicon Valley" had not yet been invented.

I worked at SRI in the early 1970s, and then moved on, returning only from time to time for brief consulting assignments. So I was surprised and happy in December 2000 when I received an invitation to attend a special reception at the International Center to honor a man who had been my boss in the early days of Arpanet, the network that served as a prototype for the Internet.

The request, appropriately enough, came through network e-mail. It read: "You are invited to join the friends, colleagues, and supporters of Doug Engelbart at a reception in honor of his receiving the National Medal of Technology. Hors d'oeuvres and wine will be served."

Appended to the notice was a hyperlink to a Yahoo! map guiding us to the building, in case any of us guests had forgotten the way to one of the world's best-known centers of computer science, mythical site for one of the very first Arpanet hosts, and stage for the most dramatic, emotional, and character-building episodes of our early careers.

The crowd that milled around the polished floor of the SRI International center that day was a remarkable mixture of gray-haired academics, bearded programmers, network visionaries, futurists, and CEOs of web-based businesses. Elizabeth "Jake" Feinler, former SRI staffer and the woman who coined the term "dot-com," was there, along with several pioneers of American cybernetics and many of the staff members I remembered from my days on Doug's team. In the front section of the room, on an easel, stood a large color photograph of Dr. Engelbart receiving the medal of technology from President Clinton. In another corner was a glass cage with a crude wooden device equipped with two wheels: the first prototype of a computer mouse. Doug spoke of his current projects, of making the totality of the web more accessible, more malleable. He spoke of the "unfinished revolution" of the use of computers by human communities.

Among the crowd there was a palpable sense of the long road traveled in those thirty years since the birth of Arpanet, mixed with anxiety at the monster that had been unleashed. Much of the discussion went beyond the party chitchat of such occasions. "What comes next?" everybody wanted to ask, looking around the room for old friends who might have part of the answer.

There was a definite sense of unfinished business. Something was wrong. The Internet of today is not the network we all thought we were building. And the way it is being deployed as a new medium for human communication and commerce raises serious questions about the kind of society we will experience in a few years.

Understanding the Internet

It is one of the ironies of history that people who live through a revolution are least likely to understand it. Nor do we realize where it comes from, where it is taking us, and where the various currents within it find their roots. I was fortunate enough to keep a personal diary and to collect copious notes during my own association with this technology. Without such a resource I never could have kept these developments in perspective.

The rise of the Internet and its multimedia extension, the web, represents such a contemporary revolution, and we are right in the middle of it. The changes it precipitates are so rapid and profound that any analysis appears obsolete even before it gets into print. The creation of the Internet is well within human memory, yet most of the people who use the network every day would be hard-pressed to describe its history. Many of the venture capitalists who fund network start-ups, and indeed many of the programmers who work for them, would flunk a simple test asking who invented the mouse or when the first packet network was first demonstrated.

In the early 1990s the Research Council of Canada undertook a major study of the future impact of the Internet, projecting greater usage of e-mail and bulletin boards, and published it just before the explosion of the web, browser software, graphic interfaces, and electronic commerce, insuring instant obsolescence for the massive official report.

The Internet and the web seem to have happened overnight. When their impact is discussed by politicians or by sociologists, it is in terms either so gloomy or so enthusiastic as to seem absurd.

Ask people when the Internet began and they will answer 1995, because that is the year when dot-com start-ups began advertising on television and widespread coverage of web-based products appeared in the mainstream media. In the following years Wall Street itself lost all perspective, hyping the valuations of baby companies to stratospheric heights, then dumping their stock like yesterday's newspaper, mere fishwrap. It was clear the financial wizards had no more grasp of what they were hawking than the ordinary folks who loaded their retirement plans with the shares of meteoric companies like Pets.com, Insweb, Dr. Koop. com, or a thousand other firms that would wither within a year.

Not only have we forgotten (or never understood) where the Internet came from and what it was designed to do, but many sophisticated computer programmers have only fuzzy notions about the history of the technology they are helping develop. There are some TV documentaries and a few popular books about the Eniac, supposedly "the first computer," and the theoretical exploits of master logician and cryptographer Alan Turing, who pioneered automata theory during World War II, but the details are only available in obscure documents couched in technobabble.

One glaring example of misleading history is the popular movie *U571*, the story of a heroic submarine crew that managed to capture a German "Enigma" cryptographic engine from a disabled German U-boat on the high seas. The movie was produced by Dino de Laurentis and released in 2000, and the narrative had been slightly touched up, in typical Hollywood fashion, to substitute an American crew for the brave British sailors who actually pulled off the feat. Another historical fact bites the dust. In the words of one reviewer, the movie is "a testosterone-fest of sweaty, gritty sailors shooting at lots of stuff and blowing things up."

As far as the general public is concerned, a few sketchy references of this kind are accepted as sufficient background to discuss the history and future of computer technology and, by extension, the impact of the web. Engineers who should know better will casually assure you that Apple invented the mouse, unless they are certain the breakthrough came from Xerox. Neither is true: the mouse was created by Engelbart's team at the Stanford Research Institute when Steve Jobs was barely in high school.

There are only half a dozen books about the Internet, written by people who weren't there when it was being implemented. It is tempting to begin a book on the subject with the old quote, "Everything you know is wrong!"

This position of ignorance in the midst of a revolution is not just ironic. It is deplorable and scandalous. It only serves the interests of marketing hacks who would convince us that the latest computer they were hired to promote, the latest dot-com, the latest programming language, holds a golden key to our future.[1]

As I was preparing this book for publication, my friend Peter Beren and I discussed these issues over lunch in a downtown San Francisco restaurant. We asked our techno-savvy young waiter to settle a disagreement about public perceptions of the web. "Are you worried about the Internet?" we asked him.

"More and more it's being controlled by big companies," he replied. "I see Microsoft everywhere when I use the Internet. And America Online."

"What do you think is going to happen?"

"They'll take over the web if people don't fight. I hope Linux wins in the end. Linux is an open system. Anyone can use it or modify it."

Another partial truth. Linux is a new operating system that is ideal for many applications. But all the big companies have understood this. IBM, Intel, and many others are already promoting Linux for their own interests. The marketing folks keep rewriting history, telling you their vision of the future of the Internet is inevitable.

They tell you this because they want you to buy the product. In the process, they try to show that the whole history of the technology is in line with their grand design.

Yet the development of the Internet was not a steady forward march leading to Microsoft, AOL, or Yahoo!. It was born out of a long series of insights, failed experiments, visions of genius, and many chance happenings along the way. Events no scientist could have forecast.

What can you do about this?

You can learn about the real history of the technology: the roads not taken, the hopes that were dashed, the alternatives that were not funded, and the safeguards that were not instituted. And you can make an impact on the future landscape of information media.

Threats and Opportunities

We are influenced by the expanding network technology that is changing everything around us—our lives, our jobs, our ways of communicating, and even our choices of community, the way we raise our children and the toothpaste we buy. Even after the collapse of the dotcoms in 2001, electronic commerce is steadily growing into a seven-trillion-dollar industry, with half of the traffic international in nature.

This truly opens up a creative universe offering extraordinary opportunities. In the developed world computer networks have created a new reality of rich communications, the sharing of ideas, information, knowledge, even emotions, in a way never before thought possible. It represents a major impact on every social and intellectual pattern in our culture. In less-developed parts of the world, the rapid changes could be so blinding as to be devastating.

Having said this, it would be a mistake to find all the benefits and all the perils within the technology itself. The potential threat is very human. It comes from the forces that are beginning to control the Internet, to bend it to immensely powerful interests, and in the process to exert increasing control over the lives and the thoughts of all its users.

In the heady, early days there was a vision of creative freedom and expanding access to information among those who created the Internet. That vision was inherent in every tool and every piece of technology they built. It was implemented the hard way, over years of trial and error, success and failure, boring research and blinding breakthroughs.

It is that epic battle for the freedom of information that today's promoters and PR executives have hijacked. The dream of the Internet's founders has been declared irrelevant and pushed under the rug of their complacency. In the process, the original vision of the network has been distorted and betrayed. The goal of this book is to recapture the clarity of purpose of the founders and build on its untapped potential.

There is much more at stake here than academic accuracy. The truth rarely matters in science, and even less in the politics of science. Second-rate figures and lucky bureaucrats routinely get credit for inventions they never made. Mathematician Bernoulli plagiarized his son's discovery of the "Bernoulli equation," backdating his own book so it appeared to have been issued before the young man's work; American explorer Admiral Peary claimed he had reached the North Pole when he knew he was hundreds of miles away. Closer to us in time, Antony Hewish of Cambridge University received the 1974 Nobel prize in physics for his "decisive role in the discovery of pulsars," but it was a young physicist named Jocelyn Bell who actually made the discovery and recognized pulsars as stellar objects!

The history of scientific discoveries gets rewritten all the time for the sake of national pride or textbook convenience, with little or no consequence for our everyday lives. In optics, Snell's law of refraction had been discovered nineteen years earlier (in 1602) by Thomas Harriot. The Chinese had invented printing a long time before Gutenberg. Nikola Tesla introduced radio before Marconi. To this day the French believe they invented the Internet because France Telecom once implemented a widely used system called the Minitel. It was actually a closed system, and the exact opposite in structure of the Internet protocol. But such matters of cultural superiority always dictate their own belief system.

What is at stake here is more important than historical truth. In some parts of the world the Internet is already being used to create a new information environment manipulated by commercial interests, monitored by governments, censored by bureaucrats, and biased to encourage conformity with the new rules of the marketplace.

Think of the ease with which the Internet provides access to public records: on the web, you can apply for a dog license without leaving your apartment. But the folks next door can use the same tools to find out in which political party you are registered. These records were always available, but one had to go downtown and waste a couple of hours to consult them. The web has changed all that, upsetting all our notions of privacy in the process.

Such breaches of privacy were not intended by those of us who were involved in building the technology of modern networks. It is a duty, for the people who carry a small part of that history, to describe the ways in which it can be misused, and to outline the consequences.

Myth and Reality of the Internet

In the first part of the book I propose to indulge in a visit to a museum of sorts, reconstructing the dusty history of computer technology and destroying a few convenient myths: The Eniac was not the first computer; the Internet is not the first network; Apple did not invent the mouse, and neither did Xerox. The opening section of this book, "The First Explorers," recounts the steps that led to the real thing.

The second section, entitled "Making the Planet a Better Place," describes the series of breakthroughs that transformed computers from calculating machines to universal platforms for new media. Much of the material in these two sections is based on first-person recollections, notes, and fragments from a diary I kept as a systems programmer at Stanford University and at SRI, and later at the Institute for the Future when I served as one of the "principal investigators" for the Arpanet. It is interesting to observe how accurately our small research community forecast the networks' potential, and also what enormous miscalculations we made about their timing, economic reality, and actual impacts.

The third section, entitled "The Betrayal of the Internet," takes us to today's network technology in the harsh terms of the marketplace, the urge by major interests to influence not only our purchases but our thoughts, and the government's obsession to harness the whole system to its own narrow definitions of security—sacrificing our privacy and possibly our freedom in the process.

In the fourth section, "How We Can Save the Dream," I state a new set of principles for network citizens and ask how we can create new standards for the use of the network.

The Challenge of Future Networks

By their very nature, networks are complex and fast-evolving. New forms of communication are being invented all the time. Therefore, it would be inaccurate to characterize the forces that act on the Internet as right or wrong, black or white, good or evil. Governments have legitimate concerns about the use of technology by criminals, and commercial firms do have a need to gather information about their users, if only to serve them better.

But there is a basic law in cybernetics stating that "information is control." These two are closely related. And the concern about privacy is only part of the issue.

In our infinite thirst for information, how much control are we ready to relinquish? In our bumbling search for better communications, are we becoming slaves to a technology that was supposed to serve us? Writer Paul Valéry posed that question in the 1920s when he took a deep look at the "modern" machines of his time and observed, "They want well-trained humans!" That was before computers, before the Internet.

Are we becoming the "well-trained humans" that a new wave of sophisticated machines has demanded? It would be ironic indeed, and tragic, if the magnificent network that was built to enhance human freedom and creativity turned out to be a primary tool of our enslavement. In the end the Internet is forcing us to ask, how can we remain human in the world we are creating?

Part One

The First Explorers

With the means of almost instantaneous communication of intelligence between the most distant points of the country, space will be, to all practical purposes of information, completely annihilated.

—Commerce Committee, House of Representatives, 1838
(commenting on Samuel Morse's invention of the telegraph)

I

Getting Out of Our Sphere

Like many of my contemporaries, I felt inspired to a life of research when the first artificial satellite was launched. I was a young Frenchman of eighteen when Sputnik I went up. My father refused to believe the announcement for several days. He was an educated man with good knowledge of math and a World War I officer's understanding of ballistics. Yet he just kept fuming about the journalists' inability to see through what he regarded as the most ridiculous Communist propaganda in years. Even when American reports confirmed that the cotton-picking Russians really had sent the orbiting curiosity into the heavens, my father's opinion was that man would never be able to live in a space capsule, much less on the moon, because "man cannot get out of his sphere."

I found this expression striking, although I could never get a definition of what "man's sphere" was. Since Sputnik, other breakthroughs have forced me to ask the question again, most notably in the technologies of information. The advent of powerful computer networks poses as great a challenge as space travel. The Internet does not simply take us out of our sphere, it redefines it totally.

It is this observation that motivates me to alert others to the potentially dangerous evolution of the Internet, and what we can do about it.

When most Americans discovered the web in the late 1990s, they thought they were witnessing the emergence of a new form of civilization, one in which entire human communities could arise around important issues, one in which solutions could be found for the key problems of our times, and where it would be fun to wander off, learn about unsuspected treasures, and make new friends everywhere.

Today, as one of my correspondents put it, "I use the net to buy books and to send out e-mail, which is great for my business, but what the web gives me is just a rehash of what I see on TV, the same depressing news, and poor-quality pictures I can get better elsewhere."

It should not be this way. Network technology is the most powerful tool available today to improve human society. Every man, woman, and child now have the ability to access facts, organizations, and—most important—other human beings. Digital networks can be used by a repressive government to look for undesirables or to flag people it considers suspects. But they can also be used by individuals to share thoughts and facts, novel ideas, visions of humanity's future destiny. They constitute communications media unparalleled in human history. And they lead us to a momentous decision.

The Choice

Networks of computers and other digital appliances like cellular phones and Palm Pilots are forcing us to make a choice between two forms of digital worlds: for convenience I call them the "Solid State Society" and the "Grapevine Alternative."

In the Solid State Society, networks are developed primarily as tools of commerce and control. Massive amounts of technology are used to support and exploit people's business activities—their daily purchases of goods and services—by clever use of advanced forms of statistics such as data-mining and monitoring of behavior. Their use for wide-open private communication such as e-mail, instant messaging, or conferencing is secondary at best. Yet the benefits of group communication through networks already were demonstrated twenty years earlier, when Doug Engelbart's group and my own team at the Institute for the Future were able to link together large conferences.

Our software used both instant messaging and electronic conversations on specific topics, from crisis management to long-range planning. The record, private or public alike, was retrievable at will, in a form of human interaction that the chat rooms of today have not yet approximated.

In the Grapevine Alternative computers are used primarily to build such personal networks. The power of this approach is stupefying: Not only can anyone on the planet find kindred spirits with similar interests or complementary knowledge, but it is possible to assemble powerful groups working for social change, artistic expression, freedom of thought, and scientific knowledge, far beyond the primitive tools of "chat" and e-mail. Such groups are beginning to use instant messaging, privacy software, and peer-to-peer computing to do away with traditional forms of control.

The dynamic force of digital networking began to be felt in obscure research organizations some thirty years ago. It is now exploding in public view. The explosion is helped by growing demand for new media, free music, access to databases and information sources. In the process, naturally, it fuels electronic commerce. But it could go far beyond such applications when larger numbers discover gateways to other minds, windows to unsuspected vistas, bridges across their loneliness, and precious understanding. Such was the early vision of those who worked on Arpanet, the hope that drove our own teams at SRI and at the Institute for the Future to propose new structures for information and to invent "groupware."

How can we make the choice between these two societies, which utilize essentially the same advanced technology for radically different purposes? First, we must demystify computers. We must strip them of the aura of complexity that technocrats like to weave around them. For that reason, this book will not talk about bits, bytes, and operating systems. Such knowledge is not relevant to what computers actually do.

Second, having demystified computers, we need to understand their history; it is only through such an understanding that we can learn to influence the technology. Any good information system can become a disinformation tool. Any powerful new technique carries with it new fears, and new pitfalls.

An Exploding Technology

I have worked with computers since 1960, beginning with the first commercial models of IBM and living through successive "generations" of hardware (the machines themselves) and software (the programs of instructions that specify the machine's work). Throughout these revolutions I have never lost the wonder and the joy that my first encounter with computers provided, but I have become increasingly concerned that we are leaving almost no trace of our activity at the human level. Our motivations, hopes, and fears were left unsaid, because it always seemed that the technology was moving too fast for us to stop and think.

Everybody was always waiting for the "next plateau" to arrive before they could catch their breath. It never happened.

There is no excuse for the enormous gap our profession is leaving in the book of history, in which it will appear that computer technology emerged into the Internet age without transition or friction from the shadows of the last world war.

The literature of computing—a science that did not exist sixty years ago—is already filling up entire buildings. But it consists of technical information, couched in the obscure jargon of bits and bytes, routers and servers, concentrators and modems, pushdown stacks and recursive procedures. This amorphous technobabble floats gently atop a sea of acronyms, and acronyms of acronyms, at the extreme edge of the capabilities of the English language, so that only the writer and the minuscule technical community around him can comprehend what is being discussed, and then only for the brief period between the time the idea seems preposterous, farfetched, and impractical, and the time it is already obsolete.

No wonder the public is confused about how to harness this technology that is encroaching on our everyday lives. No wonder we are helpless to reassert our own rights within the world it is creating.

Computer scientists have documented everything in the world except their own work. The human side of the technology is not recorded anywhere.

I know that scholars will disagree with this. On the shelves of every sociology department are ponderous tomes discussing the impact of computers on many subjects; but only an expert can decipher the statistical relevance of surveys and impact studies which, in the final analysis, have little meaning and fail to guide the reader toward practical decisions. Most of the technical information that reaches the public today is being written by the marketing departments of major firms, the analysts who follow them, and the public relations hacks they hire.

Many research reports sleep in the archives of the government, gathering dust. They, too, hide the true story of computers: Washington is as puzzled by the beasts as everybody else while strange new networks are being grown and forcibly spliced into the nervous system of the old culture. New forms of love, worship, and crime are taking shape in a social explosion that has no precedent. Again, it is largely going unrecorded unless it takes the form of a public uproar over some sensational abuse of the technology: pedophilia, gambling, fraud, absurd wealth, or abrupt bankruptcy.

This is not a new situation for a field that many writers have found too complex or too abstruse for a good story. There are precious few books about the early history of IBM, one of them officially authorized by the amazing company that has shaped so surely the technology and, through the technology, the world we experience.[2] Other personal accounts of life with computers are cautious and cold, tempered by the care taken to anger no big company and to preserve that most cherished illusion of technological PR: the appearance that the human race, good or evil, has some measure of control over its creations.

We are in danger of losing control of this exploding technology. But we can still hope to influence it, to bring a new measure of intelligence to our usage of it.

As a research scientist with the use of the computing center of several universities, and as a computer engineer with industrial companies, I have followed the technology closely. Many others had already made it their business to ponder the implications of the new machines. Eagerly, I tried to learn from them. I recall one meeting of an international standards organization at which I was introduced to a gentle lady with white hair, whom everyone regarded with obvious admiration. I was told that she was the person who had tapped the founder of American computer science, Professor John Von Neumann, on the shoulder one day to tell him that the world's first scientific computer needed a "stop" instruction. But there is no "stop" instruction for the network-based society we are now building.

If you listen to presentations by marketing vice-presidents of Cisco or AOL, there is no reason to worry about the dizzying pace of network expansion, or to waste time documenting what we do. These companies barely have time to train one generation of technicians to cable their routers or write their web pages before the product itself changes.

What could be the use of philosophizing about machines that are already obsolete when they hit the stores? Kilobit modems that deliver thousands of words per second to your home are objects of prehistoric irrelevance, ready to be replaced by megabit devices capable of supporting video streams and multimedia games. Gigabit modems—a thousand times faster—are on the horizon. The world they open is filled with dizzy new freedoms. You can exchange information with anyone on the planet, without even the cost of a stamp. You can hear music and watch movies from the ends of the Earth. New structures are being erected for human understanding. Faster. Cheaper. More democratic.

Those are heady ideals indeed. But they are based on assumptions that are too simplistic. The potential for extraordinary benefits from this technology is certainly real. But with these benefits come myths, dangers, and complex enigmas. They find their origin in the very basis of cybernetics.

Information Is Control

One of the founders of cybernetics, the late Norbert Wiener, defined it as the "Science of Communication and Control in the Animal and the Machine." This definition, as Stafford Beer has since pointed out, suggests two ideas.[3] The first is that distinctions between the animate and the inanimate, inherited from the Greeks, do not apply to the laws of regulation. The second idea is that communication is control, and that *information is control.*

Any book concerned with networking must begin with this fact.

There is no such thing as obtaining information (by consulting a file, for instance) without obtaining a measure of control over the objects or persons that the file describes; hence the fascination exerted by the Internet for advertising firms and marketing tycoons.

The meaning of Wiener's observation goes deeper still, for the love affair with computers has always been symptomatic of a quest for power. Often disguised in academia as the scholarly pursuit of information, or in the business world as "the mere compilation of passive data," the unconscious motivations of network architects are difficult to discern, and their impact almost ungovernable in the age of e-commerce.

In a book called *The Mechanization of the Mind,*[4] Jean-Pierre Dupuy revisits the early history of cybernetics and shows that technology could have a followed a different, richer path. In the 1940s the term "cybernetics" actually referred to the science of the mind, as pioneered by physiologist Warren McCulloch, rather than Norbert Wiener's mere "steersmanship." In the words of Igor Aleksander, a professor of neural systems engineering at Imperial College, London, "an opportunity for cybernetics to change the course of the philosophy of mind was missed when intentionality was misinterpreted as the providing of coded knowledge."[5]

8

Today we may be missing a similar opportunity to develop networks for human understanding rather than networks for human control and exploitation. We are providing "coded knowledge" and neglecting the "philosophy of the mind."

Today, more than ever, we need to make full use of the power of networks. Not to get out of our "sphere," but to expand it.

This was the promise of computer technology at its birth, and we have drifted away from the set of values that inspired it.

Clouded Origins

The most beautiful sound I have heard was the sound of the memory drum of an IBM 650 when the computer died. All power would go out. The motors would be still. Lights would stop blinking and, of course, the program would be lost. I would suddenly become aware of the summer sun in the dusty courtyard behind me. I would hear the birds playing and singing. But it would take ten minutes or more for the big drum to slow down to a complete stop. The high pitch gradually turned into a sustained, thrilling note, unnoticeably shifting to a hum, a rumble, then just a murmur. Eventually the drum joined the rest of the computer in death.

This kind of incident happened to us once or twice a day when I was working at Paris Observatory, because our power supply was unreliable. The year was 1961, and the machine was located in what once had been the stables of the king's mistress in the castle of Meudon. We used the machine to compute orbits of artificial satellites. The satellites went around the Earth in ninety minutes. It took our computer two hours to do the computation, so we were always hopelessly behind, even when we were lucky and the machine didn't die. When it did, I had the consolation of listening to the beautiful sound of the drum as it slowed down, imperceptibly, like the sun setting on a quiet sea. Then I had to reload the program and wait another two hours for my results.

The IBM 650 was the first electronic computer to be commercially successful in a major way. It wasn't the first computer, not even the first commercial computer. The IBM corporation, already a giant, had gone through its formative years with machines that used circuit boards for the control of their operations. You would spend hours wiring up those boards to instruct the machine on which card columns to read or which to punch. You would load the board into a sliding door and—magical moment!—hit the "start" key. The computer would sing and hum and swallow one deck of cards after another.

Even those early machines, however, were not the first electronic computers. The real origin of the modern computer is clouded in human memory and state secrets.

If you are one of the millions of people today who own a home computer, or one of the thousands of folks who will invest in a PC or an Apple personal computer this week, you are acquiring a device whose antecedents go back before World War II.

The theory of automatic digital machines is largely credited to an Englishman named Alan Turing. The practical work, on the other hand, came from two Americans, Presper Eckert and John Mauchly, who obtained the first official patent in the field. These "facts" have remained unchallenged for many years. Jerry Rosenberg's excellent book, *The Computer Prophets*,[6] for example, states that "Alan's abstract computer or 'Turing Machine' as it was commonly referred to, represented his masterful contribution to the development of the computer."

Information science students are taught that the Turing Machine is nothing more than a "thought experiment," a convenient imaginary structure to be used to prove the basic laws of automata. They are told that Eckert and Mauchly were the first to invent a practical computer, and that they went on to build the Eniac, an immense assemblage of glowing tubes and clicking relays filling many metal cabinets. The Eniac was built to solve strategic problems of World War II. Ironically, it was completed two months after the surrender of Japan.

Sperry Rand later acquired the patent rights. Lights would dim all over the nearby city of Philadelphia when the computer was turned on, and pundits predicted that seven such monsters would suffice for all the calculations the world would ever need.

This is all very recent history, and anecdotes about it abound, and you would think the facts would be straight. Yet most of these contemporary statements are false, or at least grossly misleading. It took a federal judge named Earl Larson to reconstruct the true story behind Eniac, and a British historian named Anthony Cave Brown to uncover Alan Turing's actual machines.

Some of the facts came to light when Sperry Rand sued Honeywell, charging that it had infringed upon the Eniac patent describing the invention made by Eckert and Mauchly. In his 1974 decision, Judge Larson stated:

> Eckert and Mauchly did not themselves first invent the automatic electronic digital computer, but instead derived that subject matter from one Dr. John Vincent Atanasoff.

"Doctor John Vincent who?" most people must have asked, as they rushed to their copy of *Who's Who in Science?* It turned out Atanasoff was an associate professor of physics and mathematics at Iowa State College in Ames, Iowa, where he had produced an operating model of his computing machine. In December 1939, having demonstrated the basic principle, he started work on the first actual unit, aided by a student named Clifford Berry. The purpose of their machine was to solve the difficult equations used in physics.

In an interview with the magazine *Datamation* published in February 1974, Atanasoff explained that everybody laughed when he started using vacuum tubes for digital calculations. He ignored the laughter and did it anyway.[7] He went on to build what he called a "regenerative memory," using capacitors because he didn't have the money to buy tubes. He represented numbers inside the machine using the binary scheme of zero and one because the memory cells had only two states. The machine became known as the Atanasoff-Berry Computer, or the ABC, and had no competition until 1942.

Atanasoff and Iowa State University never filed a patent for the device, which was dismantled after the inventor went to work at the Naval Ordnance Laboratory. In the meantime, Mauchly had been greatly inspired by what he had seen during his visits to Ames, and he incorporated these concepts into the Eniac when the Army funded the project in June 1943. Dr. Edward Teller, among others, used the Eniac when he worked on the hydrogen bomb in Los Alamos in 1946. A patent was eventually filed in June 1947.

The excitement of those early days can be felt in the remark by Atanasoff that "in 1939, when I looked at that little breadboard model—it fit on a desk—I realized I could compute Pi to a thousand decimal places easily enough." People who worked on the Eniac have described to me the sheer dizziness of a period when every problem opened whole new branches of technology.

The idea of the stored program (putting the instructions into memory), which originated with Von Neumann, led to the concept of artificial language, as people gradually realized that the instructions given to the computer constituted a language with its own syntax. It became possible to think of creating languages of "higher level," closer to human forms of communication, in order to simplify dialogue with the machines.

Early Computers: A Family Tree

Great science rarely appears at a single point on the Earth and in a single brain. The digital computer is no exception. Chart 1 shows my reconstruction of the major branches of development that eventually led to the modern computer. It highlights the fact that isolated teams, in several countries, were working on similar experiments, each holding a little part of the solution.

Thus, a German engineering student named Konrad Zuse, who was totally obsessed with the idea of building a computer (and had in fact designed mechanical models out of an Erector set as a child), gave up his regular job in 1936 to start work on his brainchild. He created the Zuse I, a machine that used a binary system and a primitive memory and processing unit. It was driven from a keyboard and the results were shown by light bulbs. From this prototype, Zuse went on to build his second machine using electromagnetic relays, a breakthrough that English writer Christopher Evans cites as the first use of relays in any computer system. He also replaced his keyboard with a tape system, using discarded photographic film as the medium, in which he punched holes to give instructions to his computer.

In 1938, a friend of Zuse named Schreyer was awarded a doctorate for showing how electronic tubes could be used instead of relays to create a computer memory. As Evans pointed out in a perceptive book, *The Mighty Micro:*

> For a brief period, as they discussed Schreyer's idea, there is no doubt that both men were favoured with a glimpse of the future. But inevitably, the realities of the present pressed down upon them. Valves (tubes) were rare, unreliable, extremely expensive, enormous consumers of raw electricity, and the heat they gave off when assembled in numbers would probably cause the rest of the machine to malfunction.[8]

Several years before the first U.S. and British machines, Konrad Zuse built the Z3. Completed in 1941, it used the binary number system and could perform floating-point arithmetic.

By 1943, Zuse was working on his Z4 model, which still used electronic relays and inspired a line of machines that were applied to aircraft and missile design. That same year, a young math professor at Harvard named Howard Aiken, who had managed to convince IBM to invest a million dollars into the development of an American computer, unveiled the Mark I. It was a giant machine, using relays for its logical structure, and the momentous development of computers had started. In Philadelphia, Eckert and Mauchly were already at work on a machine that would harness the speed and power of electronic tubes, and in February 1946, for the first time, the Eniac was turned on.

While the lights of Philadelphia dimmed to feed the machine that newspapers would call "the first electronic brain," equally momentous events were taking place in Great Britain.

The British Contribution

By the time the Eniac started working, several large digital engines had already been built in England, and the societal effects of the technology had become evident to the few men and women who knew about them. The story is filled with adventure and tragedy, and it involves a scientific domain that has proved crucial to the unfolding of the entire story of computers. The domain in question is cryptology, the ciphering and deciphering of secret messages, and their application to espionage of both kinds, military and industrial.

Properly speaking, the British machines were not general-purpose computers. They were designed to emulate the operations of the cipher machines used by most advanced countries to protect their state and military secrets. In 1934, the German government had changed its cipher system to experiment with a machine called Enigma, based on an invention by Dutchman Hugo Koch of Delft and originally designed for business secrets. It was a rotor cipher machine, which was pronounced secure from sophisticated analysis because the enemy could never break the code unless it knew both the mechanism of the machine and the particular keying procedures used to transmit messages.

British Intelligence learned of the existence of Enigma in several ways, as masterfully described in Anthony Cave Brown's book *A Bodyguard of Lies*.[9] In June 1938, Sir Steward Menzies, who headed up British Intelligence, received a message from the Resident at Prague regarding a Polish Jew who had worked as a mathematician and engineer in Berlin, at the factory where Enigma was built. He had later been expelled from Germany because of his religion, and was looking for a haven in France under a British passport. In return for this favor, he was willing to build a replica of the machine. Menzies decided to send two experts to Warsaw to interview the engineer. First he summoned them to his office one quiet weekend to brief them on their mission. Anthony Cave Brown describes the scene:

> The three men met in Menzies' office beneath a portrait of his patron, the late King Edward VII, dressed in tweed and deer-stalker, a shotgun in

13

one hand, a brace of grouse in the other, and a gun dog playing in the heather. One of his visitors was Alfred Dilwyn Knox, a tall, spare man who was England's leading cryptanalyst. His companion was Alan Mathison Turing, a young and burly man with an air of abstraction and a reputation as an outstanding mathematical logician.

Turing was then an assistant to Knox, who was working for the Government Code and Cipher School (GC&CS) of the Foreign Office, located away from London in Bletchley. Shortly after graduating from Cambridge, he had written a paper on "computable numbers" that would remain his most famous mathematical contribution. He proved that there were classes of mathematical problems that could not be solved by any fixed and definite process, such a process being defined as something that would be done by an automatic machine.

From Enigma to Madam

A summary of Turing's contributions to mathematical logic can be derived from several standard textbooks. What they fail to point out is that the German Enigma could be regarded as an automaton, and thus could be emulated using the principle of the universal Turing machine. Turing's friends said it would be impossible to build this machine, however, because it would have to be as large as St. Paul's Cathedral and would need more electrical power than Boulder Dam could generate.

That did not stop Turing, history's first computer nerd, about whom Cave Brown says there was "a very odd, childish side." Every night he followed the adventures of Larry the Lamb on the BBC to discuss them with his mother; at the beginning of the war he buried some of the family's money, then forgot where it was hidden.

Alan Turing's very appearance was at variance with the proper demeanor one was led to expect of English scholars. He would run some forty miles from his workplace to London and arrive at the staid Foreign Office in old flannels, with an alarm clock attached to a string tied around his waist. In spite of this weird behavior, remarks Cave Brown, there was no doubt of his genius.

The result of the Knox-Turing trip to Poland was the recommendation that the refugee from Germany be given what he asked for in exchange for the description of Enigma. It is this feat of espionage that formed the basis of the movie *U571*, in which the heroic men who captured the Nazi machine were American rather than British, Hollywood having succeeded in altering reality once again.

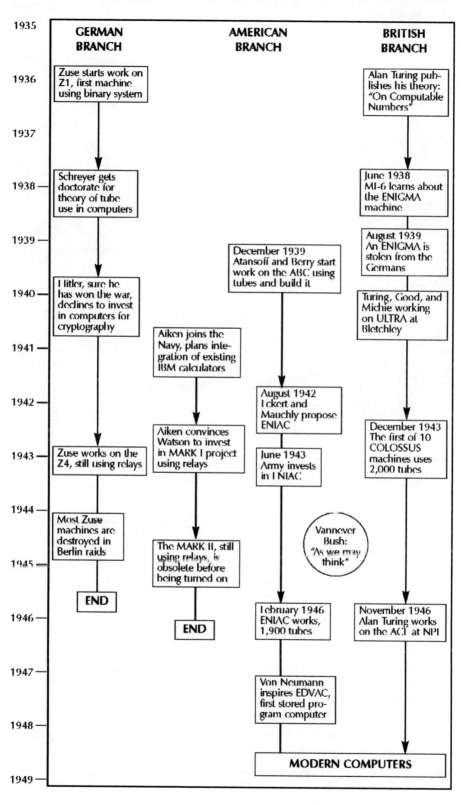

Chart 1: The Early History of Computers

Figure 1: The Z3 Computer in 1941

Several years before the first U.S. machines, a German engineer named Konrad Zuse built the Z3.
Completed in 1941, it used the binary number system and could perform floating-point arithmetic.

Once they knew the principle of the Enigma, the Turing team set upon the task of building their general emulator, which became known as "the Bomb." Constructed by a team of twelve engineers, under the direction of Harold Keen of the British Tabulating Machine Company, it was "a copper-colored cabinet some 8 feet tall and perhaps 8 feet wide at its base . . . and inside the cabinet was a piece of engineering which defied description."

Cave Brown adds that "its initial performance was uncertain, and its sound was strange; it made a noise like a battery of knitting needles as it worked to reproduce the German keys. But with adjustments, its performance improved and it began to penetrate Enigma at about the same time the Germans prepared to attack Poland."

The machine at Bletchley was useful only as long as the Germans did not know of its existence and capabilities. Accordingly, Menzies took extraordinary precautions to prevent giving away any indication that the British knew of important German moves, unless there was a credible channel through which the same information could have been obtained by conventional means; that is, without the ability to decipher the coded messages of the German High Command.

When the Luftwaffe launched a raid against Coventry, it is said that Churchill and Menzies decided not to lift a finger, even though Turing's automaton had decoded the orders deploying the bombing attacks. According to Cave Brown, they had a forty-eight-hour advance warning, which would have allowed reinforcement of the aerial batteries around the city and evacuation of the densest areas, but the British military reportedly allowed the ten-hour raid to go on.

Although other historians (such as Peter J. McIver) have disputed Cave Brown's analysis, it is a fact that Coventry was destroyed, with over fifty thousand houses being hit by bombs and only one German aircraft shot down. The British did preserve the secret of the machine that would, in the end, give them the decisive advantage over Germany. We may never know the truth about that night of November 14, 1940, but many think it marked the first demonstration of the great magnitude of the decisions the computer era would force upon government leaders.

This was also the kind of decision that would turn Turing's life into an increasingly complex series of crises. By 1943 an entire industry had

Figure 2: Alan Turing

Alan Turing (on the bus steps) with members of the Walton Athletic Club in Surrey.

been created to handle the flow of intelligence that came from the battery of his machines: about six thousand people were deciphering some two thousand messages a day at Bletchley, as Turing, who had replaced Knox when he died, began to show increasing signs of mental exhaustion.

His condition was aggravated by the stupidity of government doctors who sought to "cure" Turing's homosexuality (viewed as a threat to both morality and security) by forcing him to take drugs. Series of estrogen injections, which were supposed to lower his sex drive, left his mind depressed and his body bloated. Feeling betrayed, he was sent on a holiday by the Foreign Office but never really recovered. According to Cave Brown,

> He became progressively more eccentric—noticeably so, even in the weird world of Bletchley. Obsessed that somebody was using his team mug, he spent many hours of exacting mental work to find a way of chaining it to the wall in Hut 3 with an unbreakable cipher lock. . . . He allowed his hair to become long, dirty and wild, and his clothes were often soiled and holed.

The machines for which Turing was responsible played the decisive role in the conflict: they provided justification for the use of the first atomic bomb in 1945. The war ended, and Alan Turing was offered a lectureship at Cambridge University, which he declined. Instead, he joined the staff of the National Physical Laboratory at Teddington, England, and in November 1946 was at work on the Automatic Calculating Engine, or ACE. A working model of the ACE was demonstrated publicly in 1950, while Turing went on to work on the even bigger Madam (Manchester Automatic Digital Machine).

During 1954 he spent a weekend playing with chemicals, from which he prepared a sink cleaner and a weed killer. In the course of what started as a "game," he began manufacturing potassium cyanide and committed suicide by coating an apple with the deadly poison. Alan Turing's death is perhaps the first great tragedy of the computer age, precipitated by society's misunderstanding of the genius who had contributed so clearly to the effort to save it from tyranny.

There is agreement among military scholars that Turing's code-breaking work shortened the war by two years. A few German Enigma machines are still in existence in various museums,[10] but the British machines have not survived. It is unfortunate that Churchill, eager to hide the secrets of Bletchley Park, ordered all the items to be smashed into pieces "no bigger than a man's hand" at the conclusion of hostilities. Ten Colossus engines, the tube-based ancestors of programmable computers, were destroyed, an irreparable loss to computing history.

What did Turing himself think about the machines he had conceived and their future role in the world? We know of his views about spirituality through his correspondence with friends, as in this letter to Mrs. Morcom quoted by Andrew Hodges: "Matter is meaningless in the absence of spirit. . . . Personally I think that spirit is eternally connected with matter but certainly not always by the same kind of body."[11]

Yet Alan's main interest in the abstract logical machine he had defined led him to think of the future not in terms of a soul or life force but of intelligence. Intelligence had won the war, and as his biographer aptly remarked in connection with Turing's proposals for the Automatic Computing Engine, "his whole enterprise was motivated by a fascination with knowledge itself, in this case with an understanding of the magic of the human mind."

The early dream of the first explorers had to do with the use of computers to gain an insight into the nature of man.

Computing Research and Space Exploration

By the early 1960s, when my generation became fascinated with the challenges and opportunities of computer programming (not yet referred to as "software" in general conversation), the technology already had a colorful history. The human and social problems it would pose were detectable in the earliest applications.

Computers, by then, had firmly established themselves in businesses large and small. Many of the machines still used vacuum tubes. These machines were monsters, not only because the tubes were bulky and required massive amounts of wiring, but because they generated heat and had to be placed inside large cabinets with good air circulation. The computers had to breathe filtered air, free of dust and pollen, and the whole installation was air-conditioned. The legendary "shirt-sleeve environment" of IBM had been created; computer rooms became a favorite place for allergy sufferers.

By the standards of the computer industry, our center at Paris Observatory was behind the times. We were running a used IBM 650, a machine already on the verge of obsolescence. It had a big module that read punched cards at a time when most up-to-date machines already used magnetic tape, saving considerable time and space in the handling of data. Then there was our printer, which looked like a locomotive. Worst of all was the power supply, which came from the government-controlled utility company. Since they ran it with a reduced staff on weekends and at night, when we did most of our satellite work, the bit machine often gave up as the voltage oscillated wildly around its advertised value.

My only consolation was the opportunity to listen in wonder to the eerie sci-fi sound of the big drum slowing dying. So shrill and inhuman was the chant of the spinning drum loaded with magnetic data and tracks of hopeless numbers, I could close my eyes and imagine that I was sitting at the console of a spaceship, slowing down as it returned to Earth after a tour of the galaxy—not at the console ruling a mass of electronic circuits in the old French castle.

Computer and space exploration were closely associated. In those innocent years of the space program, when man had not yet orbited the Earth, space scientists were still classified under their original professions: they were called astronomers, physicists, propulsion engineers. Computer science was not recognized, either. Young people foolish enough to fall in love with the machines were lucky if they qualified as "applied mathematicians" (the implication being that they weren't good enough to be *real* mathematicians) or little more than engineers, which was the worst thing to be called on any campus.

I was fortunate when I was accepted for a master's degree at Lille University. It offered courses in programming as early as 1960, while

Figure 3: The Mark I Computer in 1949

Tom Kilburn and F. C. Williams at the console of the Manchester "Mark I" in 1949.

most French schools refused even to consider the subject as a part of their curriculum. Our project was something of a scandal because the faculty could not find anybody with a doctorate to occupy the programming chair. They compromised by trotting out an eminent expert in information science who came from Paris twice a week to give the introductory lectures, and they hired an IBM engineer to teach us the practical stuff. Nowadays academics will gladly boast of their big shiny computers and powerful web connections, but don't let this fool you: they gave the new discipline a very hard time before they allowed it among the "serious" studies in the curriculum.

Computer technology has been viewed with contempt by many scholars. My thesis adviser was an example of this class. He was a wonderful man and a gifted astronomer, a White Russian whose parents had come to the West after the Revolution. He had the genius for teaching mathematics that only physicists seem to develop (similarly, the best physics I ever learned was taught to me by mathematicians). This exceptional scientist felt disgusted in the presence of computers. He told me sternly that there was no future whatsoever in the computer field: "These machines are just a passing fad in science," he kept repeating as we walked along the hedge leading to the observatory. "Computers will make no long-term impact. And they waste such an incredible amount of paper!"

That was the single point that stuck in his mind. The paper. Researchers had to work with very small budgets, and paper was an expensive commodity. We would often do our calculations on the back of course material or other documents printed on one side only, if not on the proverbial "back of the envelope." So it was shocking to watch IBM machines ejecting ten or twelve large sheets of fine glossy paper just to get ready to print the next job. This was not just waste: it was a violation of the way things had always been done in science, a breach of ethics and etiquette; it announced a break in social rules. Such a machine, coming as it did from another continent, threatened a certain type of behavior that traditionally underpaid scientists had come to cherish as their own.

At Paris Observatory we had the same problem getting astronomers to accept satellites as valid research tools. Again conventional wisdom, amplified by the media, has popularized the image of the scientific community as solidly united behind the daring pioneers dreaming of exploring the solar system. Again, conventional wisdom is misleading and the record should be set straight.

Most scientists in the late 1950s and early 1960s thought satellites were just another extravagant example of military waste, with no possible application to their work. The most embittered of all were the astronomers, who should have been more excited than anybody else. The fact is that Sputnik I, launched in October 1957, not only caught the world public completely unprepared, but created total consternation among the astronomers and their calculating experts.

The Shores of Obsolescence

My father's comment about artificial satellites, stating that "man cannot get out of his sphere," was not out of line with the times. Such reactions were the rule rather than the exception among the educated public in France and elsewhere. No less an authority than the Astronomer Royal of Great Britain had said, a mere four months before Sputnik, that "space travel is utter bilge."

As soon as the news of the strange stellar wonder hit the observatories, computer teams were asked to produce an orbit for the intruder. All the experts who were "in the know," or thought they were, had confidently expected the first satellite to be launched several years later and, of course, to be American. As a result, nobody had

U.S. Army photo

Figure 4: Programming the ENIAC

Two technicians wiring the right side of the ENIAC with a new program in the "pre-von Neumann" days.

22

bothered to start studying seriously the problem of orbit computation for satellites made on Earth.

Amid the confusion that night, some bright scientist suggested that the planetary astronomers would probably have the embryo of a program that could reduce Sputnik to numbers in a hurry. Alas, the programs they did have were designed for the large distances of the solar system. They assumed that the mass of the Earth was reduced to a point. Not a very practical assumption when the satellite was a mere two hundred miles above ground, which placed it *inside the Earth* as far as the computer was concerned!

The astronomers chuckled once again over the uselessness of computers while the experts hit on another bright idea: why not ask the comet specialists? They found one fellow with a program, all right. Unfortunately, for convenience in presenting the observations, it assumed that the Earth was flat! Not a very practical assumption, either, when you were trying to track an object that kept going around the globe.

Many years later, I had lunch in Hollywood with Steven Spielberg, who was working on *Close Encounters of the Third Kind*. He was looking for a way to describe the moment when human experts would decipher the first message from approaching saucers. Their laboratory was filled with banks of computers, but somehow they had to transform the signals from space into meaningful patterns. He told me he had spent useless hours at the Jet Propulsion Lab listening to the explanations of long-haired scientists playing with multimillion-dollar machines, but he couldn't make any sense of their technical jargon. Yet the scene had to be graphic and, if possible, funny.

Recalling the early days of satellite computations, I told Spielberg that if such a momentous event did take place, it would be very unlikely that the "experts" would be ready for it. In the office of Dr. J. Allen Hynek at Northwestern University, I had seen a photograph showing three astronomers who had climbed up on ladders to fit a piece of string around a big globe. They were trying to find out where Sputnik was going, while the computer programmers were frantically attempting to figure out a more elegant solution to the problem!

The scene was recreated in *Close Encounters of the Third Kind*. The space experts of the multimillion-dollar secret team are unable to find a simple map that would tell them where the extraterrestrial landing site is situated. They end up breaking into the director's office, forcing a globe out of its precious sockets, passing it overhead down a corridor, and finally discovering that the aliens were landing near Devil's Tower, Wyoming.

The launching of Sputnik meant that a new technology had been revealed with unprecedented scope. It could be mastered only through the machines that made the adventure possible. Slowly, a new circle within the public sought familiarity with the technology of computers: classical forms of education had failed to bring it within their reach. The machines themselves kept changing, and the assessment of their impact became a moving target, increasingly blurred by its speed and by the diffraction of the haze that hid its true colors.

With each new "generation" of computers, thousands of trained technicians fell by the wayside when they could not make the intellectual transition. The mecanographers of old, who were used to punching cards, wiring boards, and tabulating results in neat little columns, had already been left behind because they could not grasp the idea of a program: "Everything you can do with a program, I can do with a wiring board!" they used to say, furiously plugging little jacks into the holes of their metallic sieves hairy with colored wires.

They were unable to follow the concept that the computer could emulate these wires in its own memory through the thousands of instructions that constituted its program, and could reassemble them in millions of combinations. Instead of moving electrical signals through wire routes, it was possible to test conditions and trigger the appropriate response, anticipating and controlling a whole universe of data.

The mecanographers did not see that the control was still there— not in hard wires they could touch, but in something elusive and invisible that the Americans had started to call "software." But to the young people who entered the field in the 1960s, software was second nature. They were a generation that was welcoming change and could keep pace with it, or so they thought.

Managers and accountants always justify the purchase of a new machine on an economic basis. Yet the real test of a new technology like computers is not that it performs the same task faster or cheaper than before: it must do something that one could not even conceive of doing before. For that reason, computers and space exploration were linked, because one was really the test of the other. The computers enabled us to get out of our sphere. You can build rockets with physical tools, but you fly them by software. When industrial life suddenly accelerated after 1960, this little distinction left thousands of engineers, in dozens of countries, stranded on the shores of obsolescence. Those who felt ready for adventure had big decisions to make in their own lives.

Neither Turing nor von Neumann, not even IBM itself, had anticipated the enormous appetite for computing the world experienced in the 1960s. Their dream had been the vision of pure mathematicians in a rarefied world of cryptography and academic research, driving forward to better understand intelligence itself and "the magic of the human mind." But that magic itself was now operating on a large scale to create new industries, to begin the exploration of space and to open up new forms of communication.

I heard the big drum of the 650 sing its agonizing song one last time. Then I resigned from Paris Observatory because nothing was changing there, nor would it ever change. I worked for a while at an electronics firm that manufactured large-scale military radars and used computers heavily in its research. We had IBM equipment and were trained by IBM engineers in well-lit offices, where everyone wore a white shirt and had a clear view of the mission. Even so, the effects were amazing, because the European industrial system has trouble coping with the rate of change imposed by the new technology. A gap, an abyss was appearing within major companies, between those who ran the computers and those who did not. I found that I wasn't particularly interested in the outcome of their feud. I decided to leave Europe, and a few months later found myself on an American campus.

Here were computers of the third generation, using transistors and printed circuits, which had replaced the banks of vacuum tubes that I would occasionally find in a garbage can outside the engineering building, now the largest building on campus. In spite of myself I was relishing the excitement of the giant machines that would, someday soon, devour their own programmers, their own children.

2

The Digital Society: Solid State or Grapevine?

The genie was out of the bottle, and the dream was changing. The search for knowledge would continue and expand, but it would now be acted out in the real world, outside the cozy confines of laboratories and computing centers. And it would feed a new passion not only to solve a few of humanity's problems but to change the world.

In the 1970s there used to be an orange road sign, shaped like a lozenge, on the drive that curves to the top of the hill behind Stanford University. On the sign one could read the words "CAUTION! ROBOT VEHICLE."

The first time I drove up that hill, to the Artificial Intelligence Lab, I was very impressed. I remember slowing down and carefully looking to the right and to the left, expecting some ugly metal monster to come down and crush my car under its caterpillar tracks. But there was no robot vehicle in view, and there hadn't been for a long time. The first model, a slow-moving chassis with three flimsy bicycle wheels and a camera "eye," had run off the road on a cloudy day. The poor thing could not see where it was going when the contrast was bad, and it got especially confused when it followed a tree-lined alley that alternated between shade and sunlight. The sign had been left there to impress visitors, particularly potential project sponsors from Washington, D.C.

A New World for Mankind?

I had friends at the Stanford AI Lab. They saw themselves preparing a new world for mankind, a world in which the old hated structures would collapse and more rational ones would be established. They saw artificial intelligence and the early Arpanet as the cutting edge of human thought, heralding new social contracts based on knowledge rather than wealth, leading to an era of universal understanding and peace at last. Was there not already an annual competition between the American and Russian chess-playing machines? Why shouldn't a rational model of our world emerge from a confrontation of the most logical tools in the leading cultures of the West and East, equipped with vast stores of accumulated knowledge?

Today, breathless projections of the impact of the web are based on similarly generous assumptions. Political leaders feel the need to visit ghetto schools, screwdriver in hand, to connect them to the Internet, preferably in full view of news cameras.

This kind of illusion is not without precedent. In the late nineteenth century, the social potential of the electrical revolution was already providing a model for human "progress." It was felt that electricity, by its universal nature, was bound to reverse the trend of concentrating power and wealth, thus equalizing social forces. In Russia, Lenin would soon propose to "electrify the countryside" as a major step toward the establishment of the power of the Soviets. In America, Chicago became the epitome of the Electrical City.

I have heard Professor James Carey point out that this vision of "electricity as social equalizer" continues today, in the concept of the networked computer as the harbinger of "universal understanding and unity." In the 1970s McLuhan announced that cosmic humanism, or something close to it, would spring from the universal appeal of electronic media. Computer hobby magazines and video games already cater to a new generation of science fiction buffs who relish the great power of their processor chips and the profound knowledge stored in their diskettes. They see the PlayStation 2 or the Mac in their bedroom or den as a personal gateway to the great networks of the modern world, and to the immense libraries whose unlimited knowledge will no longer bypass the green glare of their display screens.

They may be right. Yet I have also heard communications experts claim that "this political avant-garde which has found its ventriloquist in McLuhan is incapable of any theoretical construction; it formulates a mystique of the media that dissolves all political problems in smoke."

On this point, at least, I tend to agree with the communications experts, and I will try to explain my reasons for that agreement.

The early history of digital technology is not smooth at all. It reads like a catalog of technical misunderstandings, whether we consider the machines' applications to justice, education, government, or business. In every case wonderful claims were made in the name of progress, followed by powerful, world-changing developments at every level of industry, academic life, commerce, and entertainment. An enthusiastic technical elite predicted social or economic benefits that failed to materialize. But in every case something else happened that was indeed revolutionary in nature, something that was unplanned and remained beyond anyone's ability to control. It forced realizations that no one had intended to trigger.

The Internet and the web today can be viewed as a new environment for the same kinds of revolutions, with a huge potential for misunderstandings. We hear the same wonderful unreasonable claims and witness new, unforeseen breakthroughs for the applications that motivate further investment in the technology.

In his thoughtful essay about the failure of technology entitled *In the Absence of the Sacred*,[12] Jerry Mander observes that:

> Computer technology has sprung us headlong into an entirely new existence, one that will permanently affect our lives and the lives of our children and grandchildren. It will speed up profound changes on the planet, yet there is no meaningful debate about it, no ferment, no critical analysis of the consequences. As usual, the major beneficiaries are permitted to define the parameters of our understanding.

The Internet is changing all that. The "major beneficiaries" (media companies and large business users) are indeed defining some of the parameters, but the mere fact that the network reaches 600 million individual users means that the power does not reside solely in the hands of the corporations. We, too, can use the network in new ways to bend it to the kind of world we want to create.

So it is useful to stand back for a short while, review some of our past failures, and develop a sense of the history of such applications.

It can be described in terms of three major pitfalls.

The "Good Guys" Pitfall: Crime and Privacy

We begin with justice. The vision that placed computers in every police department and every patrol car assumed that the "good guys" would be able to fight crime more scientifically and clean our cities faster if all known criminals were cross-indexed: the felon who steals a gun in Texas can be caught in Maine. In pursuit of this vision, large computer files such as the National Crime Information Center (NCIC) were created by congressional mandate. The major build-up effort took place in the late 1970s.

Active records in the NCIC numbered 346,000 in December 1967. They grew to 744,000 by December 1968, reaching 1,447,000 by the same month in 1969. The file doubled again to 2,454,000 in December 1970, crept by another million or so the following year, and surpassed the five million mark in 1975. It climbed to *seven* million by the end of 1977, when the records also contained 132,890 wanted persons and the license numbers of more than 320,000 supposedly stolen cars.

An irreversible movement began to bring better control of crime and of the ways in which laws were applied.

The average number of transactions per day, even in that relatively innocent era, was about 260,000. Interestingly, the files also contained the names and descriptions of 17,000 missing persons who, for one reason or another, had left "normal" society, or had decided to ignore its rules and explore alternatives. With statistics like these, I cannot say that I blame them.

Today the numbers have grown even more dramatically. According to the Bureau of Justice Statistics, on January 1, 2000, there were over 59 million criminal history records in state repositories, and 89 percent of these records were automated.

In 2001 NCIC recorded 840,279 missing person reports, of which 28,765 were classified as "involuntary" (abductions or kidnappings).

What the statistics do not reveal is that this kind of tool, in the hands of law enforcement agencies, does more to change the nature and patterns of crime than to eliminate it. A brisk walk downtown around midnight, in any American city, can convince you that giant computers, the NCIC, and the web have not yet entirely eradicated the criminal element. What they have managed to do instead is to drown the whole process in the jargon of codes and identification numbers made necessary by the use of computers, obscuring even the abstruse terms of the legal profession.

Computers have also created an increased paranoia and mistrust of the "whole system," since it now appears that a suspect is facing an amazingly complex array of interconnected machines controlled by an invisible Establishment from which no human act can escape unrecorded. This, in turn, leads to the impression that no offense will ever be forgotten or forgiven by society, or at least by that self-appointed elite within society that owns and operates the computers. It also suggests to some cynical young crooks that the whole system is a joke, a sort of giant video game in which everyone, sooner or later, gets caught and is forced to play. In some environments controlled by gangs, spending time in jail becomes a badge of personal courage, proof that one has dared to defy the system, something to be proud of.

Nor is the uneasy feeling about such computers restricted to the bad guys. Legal experts and human rights activists take pains to assure us that our personal records are kept carefully segregated. Your employer should not have access to your tax records; your health insurance company has no right to know whether you subscribe to magazines on skydiving. Right?

Think again.

Any programmer who has gone through even a few weeks of basic experience will tell you horror stories that contradict such assurances. The fact that two computers are not permanently connected by an officially approved piece of wire does not mean that there is no way to create a path between their spheres of activity. Unless extraordinary security is put into place, those who have access to both of these computers can build their own temporary gateway. Similarly, when data are "expunged" or deleted from a computer (following a judge's order to purge a certain record, for example), it is generally *not* true that the data are also deleted from the backup versions of that computer's memory, kept in a secure vault in case the information ever needs to be reconstructed after a system failure. Thus, the "expunged" data could still be examined if someone really had the inclination to search for them.

As for the protection of privacy in the days of the Internet, it raises the challenge to greater levels of difficulty. It would be naive to believe that privacy will be improved by transferring a person's life history from a piece of paper in a file cabinet to the memory of a fast and powerful machine that can be reached by modem from anywhere in the world.

The "Cheap School" Pitfall: Electronic Education

If computers have had difficulty living up to their social assignment in the case of police data, their performance has been even poorer in the field of education. Yet the premise is simple: computers can take the burden off repetitive tasks; hence a single course program can be administered to thousands of kids in structured situations under the control of the schools.

As a young graduate student in the Midwest, I attended a lecture by a pioneer of the field, Dr. Donald L. Bitzer. He made some clear predictions: computer instruction would soon drop below fifty cents per student-hour. A single machine could drive a thousand student terminals. I left the lecture convinced that the solution to the educational crisis was at hand. The little red schoolhouse in the countryside only needed a few terminals and a phone line to tap into the world's most sophisticated pool of teaching resources: the memory of a giant computer, where new courses are constantly added at all levels, from kindergarten to graduate school. Students would happily go through sequences of problems specially designed and tailored to their abilities, and smiling teachers would no longer have to keep repeating the same information and administering rigid tests to students who need to learn at their own pace. All that was well known as early as the mid-1960s.

It turned out that two things were wrong with the concept. To begin with, the teachers were not smiling. They opposed the whole idea because the machines violated their social control of the educational process. They had no trouble reasserting their authority anyway, either by ignoring the technology or by showing that its proponents had not done their pedagogical homework. They "counter-implemented" the system, causing hundreds of millions of dollars to go down the drain in the process.

What made it easy for them to do so was the poor quality of the computer lessons: some of the "courseware" was outstanding, but most of the material consisted of hastily accumulated pages of information that could be found and studied more conveniently in a plain textbook. It is silly and boring to read text from a screen. There is no faster way to instill disgust among the students about *both* computers *and* the learning material. The advent of the Internet has changed some of the parameters, but the main problem remains.

The second issue was cost. When the first large-scale instruction system was marketed commercially, it was found that the company had to charge $1,000 per terminal per month. That kind of money didn't do anything for the little red schoolhouse in the country. Besides, early estimates of the time required to develop good lessons were all wrong: it took at least one hundred hours of teacher time to develop one hour of student material. The more realistic value is three hundred to one. Even then, the teacher creating the coursework has to become thoroughly conversant with the fairly complex programming language used to describe the structure of each "lesson" to the computer. In this regard the Internet is beginning to change the equation, at least for specialized sections of the education system such as adult advanced training and professional courseware, where it should prove immensely valuable. But we are still very far from the goal.

There remains one very frustrating aspect of the dilemma of computer-aided instruction: the system designers know that their concept is basically sound, yet its application falls short. Many tests have shown that some classes taught with the assistance of a computer network could be more successful than conventional classes. Students are not intimidated or inhibited by a computer the way they can be by human teachers. They can keep asking the same question without looking stupid in front of their peers and without the risk of rejection by the teacher.

Unfortunately, the machine cannot provide the student with motivation, or with advice, as a good instructor would. As a result, computer courses have often succeeded in adult education and technical training but have failed with children. Real benefits cannot be realized in our schools until the social framework of learning changes, and this calls for entirely different technologies, yet to be invented. Computers and the Internet are only beginning to provide some of the indispensable tools.

There is a pattern of generous enthusiasm for human progress in all aspects of computer development. Too often, it leads to products that are ill-adapted to the social environment they try to improve. We saw this pattern in law enforcement; we see it again in computer-mediated education.

To make effective network-based teaching happen, we need to make a much clearer distinction between the "affective" and the "cognitive" content of education. Cognitive knowledge can be delivered by computer terminals as easily as it can be delivered by books or blackboard. What is consistently missed is the affective part of this knowledge, the emotional link to what is learned. Ask kindergarten teachers to review the work of these experts, and they will tell you that an indispensable part of education is being chopped off in computer-education experiments. To correct this problem, we ought to combine human facilitation with the repetitive programs that deliver basic information. When this step is accomplished, we may see computer-aided education flourish, but this implies a new cost equation, where user support and facilitation will be expensive.

I remember my young daughter's happy eyes when I told her she could type on the bright green display I used at home to send and receive messages from friends and associates all over the country in my early Stanford days. But I asked her to wait until the telephone connection was no longer in use, because I didn't want to pay communication and computer rates while she learned the keyboard. She was very disappointed, and with all the authority of her five years, she told me she didn't want just to type on the pretty screen. "I want to type at somebody!" she insisted.

She had already associated the idea of the terminal with that of a window on the outside world, a window through which she could see her friends and exchange bright, joyful greetings. It is motivation like this that we can tap in our children to leverage into real learning.

The "Water Buffalo" Pitfall: Language Translation

With its global reach and multicultural nature, the Internet has reawakened the promise of universal understanding. One small problem remains: the six billion humans on the Earth today (soon to number eight or ten billion) speak hundreds of different languages and thousands of dialects. And the problem of rapid communication across language barriers has not been solved.

One of the earliest examples of oversell on the part of computer experts came in the 1950s, with the first attempts at machine translation. Those were the days of the Cold War, and it was deemed critical to keep abreast of the Russians. A massive effort was made to translate their technical production into English on a routine basis. Computing departments on many campuses volunteered their help to the Pentagon. Not only would computers translate the cumbersome Soviet documents into the civilized tongues of the West, but in so doing they would protect the Free World and contribute to international understanding.

That was the theory.

Millions of dollars were allocated to the venture. Then more millions. "Ah, if only we had a little more money," complained the professors to their Air Force sponsors, "we could finish the job in no time." The Air Force, excited by the prospect of mastering the enemy's thinking through a piece of machinery, kept pouring more money into the project. In the process, the government retranslated into English tons of articles by the RAND Corporation and other American think tanks that the Russians had translated *from* English in their efforts to emulate us.

Among language experts the whole thing became something of a joke. There was the story about the English text that a machine had translated into Russian, and then back into English, to be compared with the original: a *water buffalo* had been transformed into a *hydraulic ram,* and the expression "out of sight, out of mind" had been converted into a reference to an "invisible idiot." There was also a mention of a gentleman with the unlikely name of Leonardo Yes Vinci.

Furthermore, to the delight of linguists and human translators, the sentence, "The spirit is strong but the flesh is weak" had returned with the observation that "the vodka is excellent but the meat is rotten."

In a typical flurry of overreaction, the Air Force convened a prestigious panel that ruled machine translation was an impossible goal and an abomination unto the laws of language. There was a Black Paper, signed by top linguists, forever dashing the hopes of the enthusiastic programmers, and overnight the government canceled all its contracts and grants for machine translation research.

The amusing fact here is that the panel was clearly as wrong in reasserting the supremacy of the human translators as the computer buffs had been misguided in overselling their capacity to do the job, every time, and without human intervention. From the vantage point of some twenty years, we can now see that the concept of machine-aided translation was certainly valid. After all, every translator uses a dictionary, and what is a dictionary if not a machine? One man, Francois Kertecz, proved the concept's validity without fanfare and without notice.

Dr. Kertecz, at the time of the big uproar, was an information scientist at Oak Ridge National Laboratory and one of the eminent linguists on the staff on the Atomic Energy Commission. He differed in two important respects from most of the researchers involved in machine translation: he spoke several languages fluently and he was doing his work without a separate budget. When the government henchmen cut off all funds for machine translation work, they could not find him because he wasn't listed anywhere as a visible line item: he was simply part of the technical library services. So it happened that when I met Francois Kertecz, he had been translating Russian into English for years, using ingenious techniques in which the computer's role was simply to propose various choices for the translation. A human being made the final decision.

Work on machine translation resumed, slowly, under various guises, in the 1970s and 1980s. In recent years, electronics companies have flooded the toy market with primitive teaching machines that contain word dictionaries and even pronounce the word in the language you are learning. Similar services can be found on the web, to be used at your own risk. The main lesson we can learn from these machines is that the potential for both computer-aided instruction and computer-aided translation was real: they had to find their own channel into the culture. But for us website designers and multilingual users of the Internet, the massive job of bridging multiple cultures remains ahead.

Even the English language remains something of a mystery for modern software in the twenty-first century. The very latest version of Microsoft Word I am using to type the manuscript of this book refuses to recognize the word "mecanographer," insisting on replacements like "museographer." If I type the word as "mechanographer" (with an "h"), it tries, absurdly, to substitute "ethnographer." The future remains bright indeed for future generations of linguists and programmers as they tackle such unsolved problems.

The Dream Is Still Valid

Education, language, and law enforcement. I have taken these three areas as examples of applications in which information technology should be playing an essential role, three areas where the magic of computers was badly overextended and missed its target. Three examples among many that should serve as a warning when we think about the web's future.

The initial vision was valid, the dream was indeed credible, but the timing was all wrong. As an editorial in *New Scientist*[13] points out, "In 40 years of research into artificial intelligence, scientists have conspicuously failed to make a machine with anything like the comprehensive visual intelligence of a toddler. Even the most powerful computers have problems telling basic objects apart."

The technology is immature, and the design that supposedly took *everything* into account has neglected that little detail: the social structure in which it is supposed to be operating. If those mistakes have been made so consistently in the past, is there any indication that things will go differently in the future? As my friend Graham Burnette remarks, "The best minds seem to be creating new immature technologies rather than maturing what already exists."

As computers take another giant step in sophistication, they have become so inexpensive that a personal organizer like your three-hundred-dollar Palm Pilot has greater computing power than the systems aboard the Apollo capsule that went to the Moon. A corollary to this observation is that the mistakes of the past, made with slow, poorly connected machines, become magnified on our superb new networks. These systems, starting with the Internet and innumerable Intranets in the corporate world, are being developed with the same contempt for human emotional and social needs as the "management information systems" of the past.

With their power multiplied ten-thousandfold, their cost driven down, their reliability improved to the point at which mistakes can no longer be blamed on equipment failure, computers are still responding to the same motivations and ignoring the same realities they did in past decades.

Computers have not proven to be the social equalizer of the year 2000 any more than electricity was the social equalizer of the year 1900. It is only when the designers themselves recognize the effects of the technology, and reshape their attitudes toward the systems they build, that genuine benefits will be realized. I do not believe it will take very long for this to happen—perhaps twenty-five years, perhaps fifty. But in the meantime, a lot of us will have to suffer through a bizarre and painful transition, and we may experience an unprecedented form of social restructuring for which nothing has prepared us.

The dream is still valid, however, provided we avoid the trap of the easy promises made by those who simply sell the machines, the "hardware." It takes more than electricity to activate a computer: it takes a *program*. That program reflects not only the assumptions of the programmers but also the biases of their managers and the constraints of the entire society around them.

If the use of computers is often so frustrating, it is because they mirror a society that would rather not see its true face. Yet it is by revealing the flaws in the society that computers can enlighten us and propose new alternatives—just as scraping peeling old paint from an apparently ugly old house may reveal wonderful woodwork and suggest striking new color patterns. We have had the ingenuity to build this new technology over the last fifty years. Now we need the courage to plan, to decide where it should take us. That is a far more difficult and demanding task.

What Kind of World?

The expression "Solid State Society" was suggested to me by French cyberneticist Joel de Rosnay, on a dreary gray day in Paris when we were discussing the shape of future communications. It was his impression that the net result of using computers to link human beings together might not be the blissful flowering of creativity predicted by most American sociotechnologists, but on the contrary, a cold and impersonal social reality in which human contact would be minimized. He was not actually predicting such a society, but he suggested we look at it as one possible, scary scenario.

The solid-state world would be the worst form of the digital society, where each person would become locked within a personal information space with little opportunity for genuine human contact. We might even end up in some version of the world described by E. M. Forster in "The Machine Stops":[14] his characters spend their entire lives in cells underground, supported by every imaginable form of communications, controlled and regulated by numbers. The parts of the military establishment that are concerned with the survival of government in a state of nuclear conflict actually built such an underground world during the Cold War.

You cannot seriously consider the various scenarios without taking a step backward. It has been five hundred years or so since Gutenberg came up with the printing press, but that is but a small interval on the timescale of history. We have had telephones for a century, radio for seventy-five years, computers for fifty years, commercial satellites for thirty years, optical fibers for twenty years. Each of these technologies has made a change in the way business is transacted. As early as 1838, the House of Representatives recognized this when it commented on Samuel Morse's telegraph:[15]

> The citizen will be invested with, and reduce to daily and familiar use, an approach to the high attribute of ubiquity, in a degree that the human mind, until recently, has hardly dared to contemplate seriously as belonging to human agency, from an instinctive feeling of religious reverence and reserve on a power of such awful grandeur.

When we read the above lines, it becomes obvious that in observing the new computer-based media we are witnessing the long-term effects of a revolution that never stopped. After the telegraph came the telephone. Simple as it seems to be, it has shaped our lives in peculiar ways that the giant telephone company prefers not to elucidate in too sharp detail. In a newspaper article entitled "Telephones in the Country," published about 1900, a writer named Wilbur Bassett had already noticed important social effects of this novel form of communication:[16]

> A quiet revolution is taking place in Western country life, which promises to accomplish results within a year more important and far-reaching than any since the advent of the transcontinental railroads.

> Already the pioneer life of the isolated farmer has disappeared and the tide of industrial and education advance has swept over the Northwest. . . . The telephone does away with the seclusion of rural life, binds together scattered communities, creates social interests, and destroys the barriers between city and country. Henceforth the country is but a vast suburb, in touch with the metropolis of its neighborhood, unified by the voice of one leader.

Bassett even foresaw some fundamental changes in economic roles, since the farmer would be able to keep in touch with the market and "dispose of produce directly to the city dealer or to the consumer without the assistance of any middleman." Similar claims are now being made about e-commerce, with equally dismal results. Many of the unfortunate dot-com start-ups that vanished in the collapse of the Internet economy since 2000 were trying to "disintermediate" human experts: "Who needs real estate agents? A simple database will give you your new home," they claimed. "Get rid of car salesmen, insurance brokers. You can do it all automatically on the web! Bypass the middleman and get the best mortgage rate. . . ."

A hard lesson has been learned as these companies failed, but we cannot predict how new communications systems will restructure the economy.

We may wonder that so little attention has been directed in the last eighty years at an area of social research that seemed so rich.

Of particular interest is the question of control. Is it always true, as Bassett implies, that increasing communication provides the individual with a better chance at economic and intellectual survival, and with more control over his life? Or does it lead to the opposite result, greater uniformity and a concentration of control in the hands of even fewer people?

Our society is answering the question: it is the latter. Modern man, with all his telephones, his access to media and Internet connections, feels less and less in control of his environment. The most striking example of this alienation was found in the very cradle of web affluence, the San Francisco region in the year 2000, where the influx of dot-com companies antagonized large parts of the social fabric, threatened to decimate the artistic community, and drove thousands of lower-income people out of their homes.

As for the social history of the telephone, we have only anecdotes, and those are not even systematically arranged. Of its psychological implications we know nothing. What subjects are discussed today that only the phone made possible? What opportunities are lost, and more important, what psychological types are we favoring and promoting by making the ability to use the phone one of the tests of success? What cultural patterns have we created or changed? And what do we know about those people who refuse to subscribe to a telephone service, or use the device only to call others but do not pick it up when it rings? The advent of cell phones has only magnified these questions.

Expressing thoughts through a phone is not always as simple as we assume. Witness the difficulty the phone companies had in their early years to get businessmen to accept the device. In a newspaper article published about 1900 by Angus Hibbard, general manager of the Chicago Telephone Company, I find the observation that "the man who knows how to use a telephone properly is comparatively a rare personage." The article goes on with some technical pointers ("the lips should not be an inch away from the rim of the receiver and the voice should beat squarely upon the drum to which the little sound hopper leads"). It also recognized that lack of mental focus was the real trouble in electronic communication. What Hibbard said on this subject eighty years ago is still valid for our most advanced electronic media:[17]

> If your thought is not concentrated on the transmission of your message you will not make yourself heard or hear what is said to you. This is where a failure to realize that you are holding actual conversation is apparent. No person understands this phase of telephonic trouble better than the operator of long-distance lines, where conversations are important and comparatively expensive, and time is limited. He knows that, in case the two on the line do not readily hear each other, he must make each realize he is not talking into a hole in the end of an iron arm, but speaking into the ear of a man.

Some early users of the telephone resisted the device because they felt it created a mental barrier between the two speakers. This problem was overcome only when it was understood that the voice at the other end of the telephone was that of a human being.

In Western countries a new generation grew up with phone use as second nature, but the same does not apply to most nations. In the world of the early twenty-first century, a quarter of the Earth's population has never made a telephone call, and observers of technology like to point out that Africa has less international bandwidth than the Brazilian city of Sao Paulo. Even in that situation, however, the Internet can help people run their lives more effectively, whether they are fishermen in the Bay of Bengal getting critical weather forecasts from U.S. public websites or physicians in Bangladesh reading online versions of medical articles published in expensive journals and unavailable in their libraries.

The key to making the communication effective and useful lies not in the technology only, but in the human factors that surround it and condition its application.

Where Is the Human Touch?

The Internet has some way to go before it acquires the same human touch as the telephone, in spite of all its chat rooms, bulletin boards, and newsgroups. The development of sophisticated computer applications in business management, education, or artificial intelligence too often takes the form of blatant overselling of a fragile technology camouflaged under layers of false intellectual veneer. That such a process could lead us to the more evil aspects of a Solid State Society is a clear possibility as the technology finds its way into the life of the average family.

The increasing standardization of the information that people use to guide their lives can bring with it increased uniformity and control. An environment can be created that makes human contact a transient, unemotional affair. There would be little commitment to others in this electronic world. Values and attitudes would be encouraged that would decrease people's attention spans even below today's limited ability to pursue a given goal.

In an extreme form of the Solid State Society, our civilization might become little more than a giant, clean, well-ordered Disney World, where presidents are activated by invisible machines, get up and make predictable speeches before well-behaved citizens whose every possible move has been anticipated; a world where the flow of the crowd from a fake jungle adventure to a phony Native American tribal dance is monitored by machines, with meticulous obsession to ensure that enough toilet paper, soda, and security guards will be present along their path. The computers could also dispatch enough actors dressed as Mickey Mouse and Pluto to give every man, woman, and child an equal chance to shake hands with his or her favorite myth.

Hasty overreaction by American lawmakers to the threat of terrorism may well precipitate forms of the Solid State Society, but advocates of law and order are not the only ones relishing such an environment. This is the world that marketing experts and entertainment companies dream of, a world where corporations could keep track of individual customers' habits and tastes while governments could detect anything they consider as deviant thoughts, subversive ideas, or suspicious moves. In the process, much that is precious to our society would be threatened—not only art and literature, but scientific progress and the freedom to create, the ability to explore uncharted provinces of the mind and new social behaviors.

This is the world that terrorists could force us to inhabit as personal liberties are taken away in futile efforts to protect us. Indeed, in the days that followed the attack on New York and Washington, the U.S. government moved quickly to implement measures restricting the ability to use encryption and anonymity on the Internet, and enhancing law enforcement's ability to use the network to snoop on individuals.

It doesn't have to be this way.

Grapevines: Defining the Dream

No matter what we are told by government officials or business pundits, there are many alternatives to the Solid State Society. I like to think that the Internet will be used to support the diversity of cultures and lifestyles that I find stimulating and wonderful on this Earth. Far from reducing these cultures to their basest elements and dictating their paths, or falling into the trap Jerry Mander aptly calls the "cloning of cultures," the web should be a tool furthering their diversity and ability to invent and survive. It should become the new medium through which creative electronic grapevines can grow around the old systems.

What does it mean to have an electronic medium that transcends time and space? The telephone permits at most three or four people to have an intelligent conversation, but it generally breaks down after that. Television is a one-way medium. Videoconferencing is useless in practice when more than two rooms are linked. But the Internet is open-ended and essentially free. It has few limitations in terms of reach, number of simultaneous persons accessing it, or delay in transmitting critical information.

In the reverse situation of most technologies, it actually becomes more useful under increasing loads, and it permits the retrieval of facts and ideas that have once been recorded.

The dream of the builders of the Internet was nothing less than the design of a model for a universal information medium, a common pool of knowledge accessible from anywhere and anytime. It would not be controlled, as television is, by a few studio managers. Nor would it fall under the same rules as radio or telephone, which can function well only in a highly regulated environment in which frequency ranges are sold to the highest bidder and where monopolies control the flow of messages.

The power these technologies have unleashed has revealed new problems, in part because it has attracted the appetite of politicians, Hollywood, major businesses, and advertisers. These big users have a legitimate interest in providing content and structure for the new medium, but their participation should be balanced against the rights of individuals to express their own views and to build their own communities.

We may hope, before we become trapped in a solid-state world, that we will have understood why truly great thoughts seem to arise in solitude, and sometimes in sorrow. Some of the most important advances in technology, including the Internet itself, certainly did.

3

Arpanet Genesis

The first motivation for building sophisticated networks of computers came with the realization that access to information could be made independent of the old industrial hierarchies that controlled it. The inventors were not social revolutionaries trying to return power to the masses. They simply reasoned that the new structures would be cheaper, faster, and more effective at doing everything the old ones had done, whether in telephony, command and control, or the management of complex industrial operations.

Once the first networks had appeared, a second major step became possible, just a wild idea in the minds of a few visionaries: It should be possible to use the systems not only to reach information wherever it might be, but also to tap into the ideas and the facts in the minds of those who owned that information and understood it best. In other words, it should be possible to build electronic communities of experts, invisible colleges of kindred spirits.

California played a major role in both of these essential steps in the implementation of the wild network dream for reasons that have to do with its unique intellectual climate and its ability to take big chances with lifestyles, intellectual challenges, and the early adoption of technology.

The California Fascination

A visitor to Santa Clara in the 1950s would have immediately shared the Franciscan padres' delight with the beauty of the area. From San Francisco Bay to the charming coves of Santa Cruz, luscious hills unfold the majesty of the redwoods and drench people with the fragrance of eucalyptus and pine. There were orchards and flower beds everywhere. The rising sun put rainbows in the droplets of irrigation systems on both sides of Highway 101 all the way from Palo Alto to San Jose.

A strangely mystical fascination has always been felt here. Is it the ominous presence of the big earthquake fault, which has juxtaposed utterly different geological terrains? Is it the whimsical mixture of palm trees and evergreens? Is it the feeling that this is the last row of hills in the Western universe and that the blue sea they overlook is perhaps mankind's greatest remaining challenge, full of untold history and sunken treasures?

All of this is reflected along the bay in weird structures, tales of mystery, and annals of crime.

In San Jose, the Winchester House rambles on acre after acre, architectural testimony to the spiritualistic beliefs of its owner, the widow of the firearms tycoon. In Palo Alto, Frenchman's Tower is still an object of puzzlement to local amateurs, its history never fully elucidated. In Santa Cruz, a "mystery spot" seems to be nothing more than a tilted house that confuses one's sense of balance and perspective, but it brings excitement (supported by good advertising) to visitors awed by its purported violation of the laws of nature. In Redwood City, a group that believes in faith healing meets weekly, while San Jose serves as international headquarters for the vast Rosicrucian Order with its library of esoterica and a museum boasting mummies.

One can find such wonderful oddities in almost every city. But it is their juxtaposition with the new alchemy of semiconductors, biotechnology, and interplanetary research that is striking in Silicon Valley. From the dreaded Devil's Slide to the earthquake-molded hills, where the ghost of a well-meaning mystical commune once called "Holy City" remains hidden in the woods above San Jose, there is the ever present evidence of a strange past and the expectation of an even stranger future.

California is also the land where terrible earthquakes can strike without warning. If you are a young engineer with an idea, you don't wait for your boss to give you permission to implement it. If you don't act right away, you may never have a chance to find out if you were right or not. This sense of imminence and urgency provides the ideal setting for intellectual and spiritual breakthroughs.

The Stanford Miracle

In Palo Alto, the Stanford family built a monumental entrance to its grand estate: on either side of the gate were golden griffins on stone pedestals. The estate was later turned into a university; scholars came from around the globe to study and teach. They began to transform the peninsula through their work and the activities of their students.

It was in Palo Alto, and with financing from Stanford, that an engineer from the Federal Telegraph Company named Lee de Forest developed the vacuum tube, basic to the electronics industry. Other FTC engineers went on to create companies like Magnavox and Litton Industries.

During the 1930s, Stanford professor Frederick Terman, who taught engineering, advised his graduates to start their own electronics companies and keep them in the area. So when a young man named Bill Hewlett designed an audio oscillator in 1938, Terman helped him form a corporation with David Packard. Hewlett-Packard got started with a $1,000 loan from Crocker Bank and an order for several oscillators from Disney Studios. Today it is still headquartered in Palo Alto, and has become the world's largest manufacturer of electronic laboratory instruments, as well as one of the biggest computer companies.

After World War II the pretty orchards and quaint wooden houses began to disappear as buildings of concrete, glass, and steel rose all around the old Stanford estate. Then Stanford professors teamed up with the Varian Brothers to create Varian Associates. The year was 1948, and the development of the area was just beginning.

It was Terman's imagination, notes one author in a detailed study of the region,[18] that enabled the Peninsula to become a world capital of high technology:

> We have been pioneers in creating a new type of community, one that I have called a "community of technical scholars." Such a community is composed of industries using highly sophisticated technologies, together with a strong university that is sensitive to the creative activities of the surrounding industry. This pattern appears to be the wave of the future.

Terman became dean of the Engineering School, then provost of Stanford University, which leased part of its vast real estate holdings to Varian and Hewlett-Packard in 1951.

Three years later, the board of trustees announced that it would develop an industrial park for other companies, and it imposed its architectural standards on the whole area. Attracted by the beauty of the land, its well-planned future, and the ideal climate, young engineers came to the Bay Area in larger numbers. Stanford spin-offs now included the Shockley Transistor Corporation, headed by Bill Shockley, who had coinvented the transistor in 1947. Many scientific experts thought the transistor was a useless laboratory curiosity, but a few investors in the Palo Alto area followed their own gut feelings and prospered. Several Shockley engineers soon left the company to start new work at Fairchild, which became a major manufacturer of semiconductors.

The story continued: Fairchild got bigger, and tensions developed. Scientists left to form their own firms: Rheem, Signetics, National Semiconductor. In 1968 Robert Noyce and Gordon Moore left Fairchild to form Intel. Lester Hogan was picked to take direct control of the semiconductor operations at Fairchild. Jerry Sanders, who headed marketing, left that firm to form Advanced Micro Devices. Along with these dynamic engineers and businessmen were the silent actors, the venture capitalists, the bankers. Many financial institutions got the message and established their own branches in the area.

Today the orchards have disappeared, replaced by condominiums with swimming pools in dormitory towns such as Cupertino and Sunnyvale. Housing costs have risen so fast that even the most affluent computer companies can no longer afford to relocate the engineers they used to hire from all over the nation. In summer, a layer of brown smog frequently hides the hills from view. A worried astronomer from Lick Observatory, in those hills far above San Jose, told me that the staff must wash the telescope mirrors twice a year and resilver them every two years. In most parts of the world, such intensive maintenance is unheard of. California astronomers feel the drastic effects of pollution even at four thousand feet.

Augmenting Human Intellect

About 1970, at the time when I moved to Silicon Valley, the most advanced computer research project was located at SRI, the Stanford Research Institute (now known as "SRI International" following its administrative separation from Stanford University). The project was called ARC, for "Augmentation Research Center," and it was run by its visionary leader, Dr. Douglas Engelbart, whom everybody called Doug.

It was just one of many projects at SRI, with no anticipation of that day in December 2000 when Engelbart would celebrate his presidential award. Yet the goal of the project could hardly have been more ambitious: its stated objective was nothing less than augmenting human intellect.

Doug had found himself, at the end of World War II, thinking about his dead friends as he surveyed the smoldering remains of what had once been called civilization. Why were the people of the world engaged in such wanton destruction? Why did technology always seem to end up as a tool of war? The atom bomb, the aircraft, the radar, the computer—they were all serving the high priests of Mars. Why couldn't anybody turn technology to serve the good side of humanity?

Like millions of dazed people everywhere in 1945, Doug Engelbart felt sure the catastrophe he had just witnessed had been useless. His training in engineering (he had a Ph.D. in control systems) told him there must be a way to help people communicate without conflicts. There must be a method, a path, through which people could understand their neighbors, and if it took a machine to do it, then by God, he was going to build that machine.

There were other scientists who shared Engelbart's acute frustration. Vannevar Bush, in a magnificent flight of inspiration, had already proposed to create computers that would act like personal assistants to scientists, thinkers, leaders in all important human organizations. The machine he proposed would be called the Memex, for "memory extension." Doug embraced the concept and took it one step further. He decided to design a structure that would do nothing less than improve the human ability to think. As soon as he got out of the armed forces, he began looking for a place where he could build a system to "enhance human intellect." He found the place in California, in the high-tech crucible of SRI.

Augmentation Research

In Silicon Valley technical obscurity is a mortal sin. SRI-ARC may have been meant to bring peace and harmony to mankind, but the project was stuck in scientific backwaters. Doug had much difficulty explaining what his invention was and what it did. Writing was a challenge. Like many brilliant inventors, he thought in images and notions, structures and frames, but not in words. He wasted a lot of time talking to academic and scientific groups, and finally discovered that the only people who could consider his ideas for the enhancement of human intellect were in the military. But they did not understand the need for an entirely new form of technology that would link people together, so his proposals remained at the bottom of the pile on the desks of their managers.

It took the efforts of two people, whose names unfairly have remained obscure, to obtain the initial funding. At SRI a clever vice-president named Roy Amara (later to become president of the Institute for the Future) helped Doug write his ideas in a more clear, structured form, improving the presentation of his proposal. At the Air Force Office of Scientific Research, a brilliant woman named Rowena Swanson became an internal champion for the project, making sure the documents from SRI landed on the right desks and stayed on top of the pile. In the end, Doug's concepts, born in an effort to eliminate the basis of war, were funded by the Pentagon. The work would suffer from an identity crisis throughout its stormy history.

Even after it funded him, the military did not seem to understand what Engelbart was talking about any more than the academicians did. The situation became even worse after Rowena Swanson's departure from AFOSR. But the military had money and bold ideas: if Doug could come up with anything even approximating the kind of information manipulation he was dreaming about, they were sure they could use it in the new electronic world they faced. Doug assembled a good team, which wasn't hard to do in the profusion of talent thriving in the region. In a few years he created an extraordinary machine for the exploration of the world of thoughts and data.

In 1968, when it was publicly demonstrated for the first time before the Fall Joint Computer Conference in San Francisco, Engelbart's system was capable of doing on a small scale all the basic functions now regarded as the building blocks of Internet groupware. It processed text, structuring it into paragraphs and sentences that could be moved around at astonishing rates. It could merge the blocks of text into larger entities or split them into individual "files."

Engelbart at the console was linked to his staff by a microwave channel, so that he could project on a giant screen both the text from his computer terminal and the contents of the computer files thirty miles away. His assistants were also shown on the screen, editing sentences in memoranda, moving commas around, formatting entire books, and even drawing graphs they could interactively change. And superimposed on those graphs could be the faces of anyone discussing these changes with Doug. The group even kept a running "journal" of everything it did. It was the first online community, the first use of groupware.

You could not replicate what they did today, even on the most advanced version of the Internet. For one thing, you couldn't merge live video with text processing.

When Doug Engelbart began his work at SRI, the digital display so popular today was not generally available. He had to improvise. He invented a way to display text on small oscilloscopes, where it was picked up by a TV camera and reproduced in black and white, like a page of newspaper, on television consoles throughout the laboratory. The ARC team also created a way to point at particular areas of the screen. A device shaped like a mouse was attached to the table supporting the screen; if you pushed it in any direction, an arrow on the screen moved in that direction until you found the place where you wanted to tell the computer to change the information.

As early as 1968 the ARC team had demonstrated that its work was many years ahead of its time. Somehow, instead of blossoming into a major industrial effort, it quickly got into trouble with the information community and with its sponsors. The information experts seemed dumbfounded and a little jealous. Those were the days when information science conferences were devoted to discussions of indexing, catalog-sorting algorithms, and the introduction of the microfiche into academic libraries. The ARC system seemed to have dropped from Mars.

Engelbart was seen as a loner following his personal vision. He didn't feel any need for his colleagues' help; he rarely even went to their meetings. His sponsors, on the other hand, failed to see how the military could use the system. Doug could have capitalized on his advance if he had turned the ARC to something practical, if he had recruited assistants who understood it and could sell it as a product. But Engelbart was wary of such people. His research was only beginning. He preferred to be surrounded by engineers like himself. SRI-ARC became a rocky island in the turbulent flow of information technology, and the vision had to be compromised in order to survive.

At the reception honoring Doug for receiving the national medal of technology, a sense of irony was savored by those of us who had known him during that period. The people who were singing his praises and hailing him as a pioneer had spent the previous quarter century trying to kill his project.

The neglect suffered by Doug's project would be just an anecdote, to be told and forgotten, were it only a case of a small group of technicians having fun with cutting-edge research and preserving their livelihood by sheer strength of their talent. But these technicians already had created and demonstrated a microcosm of a future world.

They had anticipated the environment of things to come. The crises they lived through were prototypical. I observed them as such, realizing that their compromises prefigured a network-based society that would not be attained by the rest of the population for another ten or twenty years. The ARC story should be taken as a serious warning about the kind of community we are starting to create deliberately around the World Wide Web.

What was so profound about ARC? It sounds simple on paper: a computer program designed to facilitate document preparation, allowing people to archive and share their ideas through a running journal. It was called the "NLS System" for "Online System," another example of Doug's innocent affliction with RAS ("Redundant Acronym Syndrome"). In my view, the most important point about NLS was the potential for creating a community that could work from many different places while participating in the same creation process. This idea is beginning to blossom today, yet the morning commute still dominates the lives of millions in every city of the planet, and much of one's existence is spent looking across a desk at some other desk, and through a glass door at other glass doors.

Engelbart's vision was of instantaneous exchanges over wires of intelligence, of a pool of information to and from which, in the New Age verbiage of the mystical 1970s, "everyone could bow and drink deeply." But perhaps he did not recognize that other truth: that the technology was far ahead of humanity's comprehension of it, that the

SRI International

Figure 5: The SRI Console in 1968

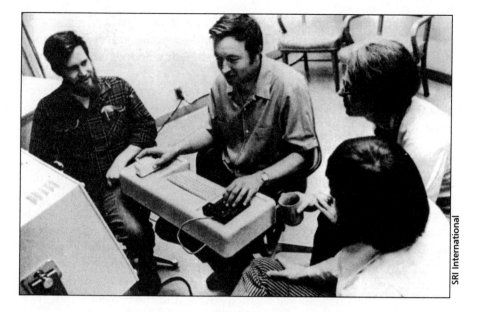

SRI International

Figure 6: Early Version of Engelbart's System

human race could not yet cope with social and psychological transformation implied by the electronic interconnection of thousands of people. Like most engineers, Engelbart and his staff underestimated the adaptation, the human factors. And the human factors came back and took revenge.

Competing Interests and Human Factors

When I joined SRI-ARC in 1971, Doug was looking for people who had built real information systems and acquired some bruises in the business world, a world he regarded with a mixture of fascination and contempt. After all, he had built himself an artificial business world of his own, seemingly immune to the storms of the external scene. He had populated his project with individuals handpicked and molded to suit his dream.

All was not completely smooth in paradise, however: the chief engineer had resigned, and most of the early systems programmers had moved on to other projects. They had been replaced by technical wizards with a strong counterculture bent and sharp views about an Establishment they regarded as utterly bankrupt. In the fast-moving world of what we jokingly called "Silicon Gulch," where suicide and divorce were alarmingly frequent among the educated, affluent, middle-aged employees of the electronics industry, the kids dropped out, learned programming, and fell in love with technology and computers. It was the only power trip in town, unless you liked to fly Army helicopters and carry an M-16 rifle through Asian jungles.

SRI recruited dozens of these young people as programmers and secretaries, because they were content with the low salaries necessary to keep overhead down. They looked at Doug with awe: he could give them a glimpse of a different future as well as function in the old world of the Establishment. Even grass and LSD seemed to be connected to this: the dream of an information world extended the exploration of consciousness that the drug culture had begun. Here, too, I thought I saw a preview of a society in which the chemical control of certain moods would parallel a supertechnology for the manipulation of information. But there was another current at SRI, and it pulled the whole project strongly in the other direction. Grants were getting harder to come by, and the project could find no sources of support outside the military.

Along came a new sponsor called ARPA, and the project fell deeper into the arms of the military establishment.

ARPA, the Advanced Research Projects Agency, had been created in 1958 by the U.S. government to react swiftly to the Russian threat in space. Following the success of Sputnik I, it cut through Pentagon red tape to speed up the launching of America's spacecraft. Soon it began to turn its attention to other areas, becoming increasingly entangled in bureaucracy in the process. By the time it looked at computers, ARPA was up to its ears in the same red tape it was supposed to be fighting. Witness one of my travel vouchers, in 1974, when my own project's accounting classification was listed as:

9861611.2411 0627 A5X09 0622 0632 0642 0652 T50219 AFG 124

I had no idea what any one of these digits meant, but I suspected that merely by mistyping a "5" for a "7," a tired clerk would have designated a Sherman tank or a ton of rubber bands instead of a computer researcher from California. Engelbart now had to guide his project into that brave new world. He was like the skipper of a sailboat with a crew of carefree, slightly inebriated weekend sailors, caught in the summer maneuvers of the Seventh Fleet.

Packet Switching and Julius Caesar

There are many "fathers" of the Internet: Bob Taylor, Bob Kahn, Larry Roberts, and Vint Cerf all have a legitimate claim to the title; and Professor J. C. R. Licklider of MIT, later a department manager at

ARPA, deserves a great part of the credit. But the names of Paul Baran in the United States and Donald Davies in England are widely quoted as the "grandfathers" because they independently came up with packet switching, the scientific breakthrough that consists in breaking up the data that move over the network into small units sent along different paths and reassembled at their destination.

Writing in the November 8, 2001, issue of the *New York Times*,[19] Katie Hafner observed that another computer scientist from UCLA, Dr. Leonard Kleinrock, who did seminal work on communications architectures, claimed to have "created the basic principles of packet switching" before Baran and Davies. Indeed, Kleinrock had published an important paper on networking as early as 1961 but Dr. Davies observed he found no evidence of packet switching in it.

The argument about who should get the credit is likely to go on for a long time; and as is often the case in science, multiple authors and multiple motivations were undoubtedly at work when the idea of packet switching was germinating. Paul Baran's own experience is a case in point.

Figure 7: Paul Baran

Paul Baran is an American engineer of Polish origin whose parents emigrated to the United States when the combined madness of Adolf Hitler and Josef Stalin threatened their country. In 1959 he joined Rand Corporation in Los Angeles, one of the think tanks that worked for the Pentagon. The problem of the survival of command and control systems under a nuclear attack absorbed him almost immediately. It is to the quiet work of this discreet man (even today he passes unnoticed at most conferences and hates to give speeches) that Arpanet, the ancestor of Internet, owes much of its early development in the 1960s. At Rand Baran invented a scheme he called by the awkward name of "distributed adaptive message block switching." Today it is known simply as "packet switching" and forms the technical basis of all the computer networks in the world.

In classical "circuit switching" practiced by phone companies, a single failure can prevent end-to-end communication. Paul Baran studied the effects of redundancy and reached the conclusion that a survivable network could be built better by using less reliable, but more redundant paths for information. If you have to anticipate heavy damage to your communications infrastructure (as you would in case of all-out war, or in a network of great complexity) it makes sense to have many possible paths for the data, and to allow each node to dispatch the information in the form of "packets," sending them to the nearest node with as little delay as possible. It's like playing a game of "hot potato": if your first recipient is busy, toss the hot potato to your second choice, and so on.

The concept was hard to accept for traditional phone companies and even for the military, where the transition from analog to digital forms of communication was still in its infancy. As Paul Baran put it,

> This fundamental difference may seem obvious and even trivial today. But its statement tended to generate an undue number of livid words from otherwise competent communications transmission engineers. Those not versed in the digital computer art tended to excessively strong objections. And most of those whose day-to-day occupation was caring for telephone lines thought that I must be crazy; a complete fraud who didn't understand how a telephone worked, or both.[20]

Packet switching gave you the ability to build extremely versatile, indestructible networks of great complexity, but it brought other benefits as well: you could now link together computers of different sizes and makes. In a remarkable interview with Stewart Brand published in March 2001,[21] Paul Baran confirmed that surviving nuclear war was only one of the concerns that led to the use of packet switching. Bob Taylor, who was head of the Information Processing Techniques office at ARPA, independently had the idea to link a single terminal to two different computers in a network. This idea led to the first notion of the Arpanet. "The method used to connect things together was an open issue," Baran recalled. The information about packet switching was already in people's minds. British scientist Donald Davies had independently arrived at the same idea. But the origin of packet switching, according to Baran, was indeed rooted in the Cold War.

During the 1950s and 1960s the command and control systems of the American military were based, like all telephone networks in the world, on communication centers linked by dedicated lines. Such systems inherited from the Second World War could not have survived a direct nuclear attack because they were built as closed networks directed from a single nerve center. Take out the nerve center and all communications stop. (This also illustrates why some public computer systems, like the French Minitel, are not really forerunners of the Internet: A single well-placed bomb could have taken it out of commission. The Minitel's structure is actually the exact opposite of the Internet.)

Paul Baran had started from the principle that the key to sturdier architecture was redundancy, founded on communication structures that would work as coherent entities even after the destruction of many of their components. This thought had drastic consequences because to build such systems you had to give up the notion of a central point of control. No wonder the telephone engineers of the older generation had trouble with the concept!

Paul Baran kept explaining his idea, from the laboratories of AT&T to the briefing rooms of the Pentagon. To achieve redundancy, you had to give up the notion of using analog signals, because they degrade with each new connection. You had to use data in digital form instead. Next, it was necessary to distribute information to the network as a whole without concentrating it along privileged channels or unique circuits that risked being cut off at any moment. Baran went one step further. He proposed to break up the messages themselves into small, independent units—the "hot potato" idea.

Slowly, the breakthrough concept of the "packet network" gained recognition. At the point of origin, a message would be chopped up like salami into small units (the individual packets) and the network would be flooded with them. At the other end, the addressee's computer would reassemble these packets, no matter how they had reached it, and the entire message would be restored. If some elements of the network (computers or communications lines) were destroyed during this operation, transmission might degrade and demand a longer time, but one would still end up with the information. Neither an outside enemy nor a natural cataclysm could interrupt the flow of data.

To understand the genesis of Arpanet and its impact, one could make a useful comparison with the works of Julius Caesar. The emperor had constructed the Roman highways to allow his legions to move quickly throughout Europe in order to crush the uprisings of "barbarians." Centuries after his death, the barbarians had won over Rome, but those roads were still used in common by merchants, peasants, pilgrims, and learned scholars. My father once showed me the remains of such a *via*—large, flat stones in a muddy cabbage field near Cergy-Pontoise. Caesar had died without realizing that by paving the major European axes, he had contributed to inventing international commerce.

Similarly, the Pentagon, which created the first big international computer network while massively adopting the "packet switching" concept to ensure the survival of its control systems, could not foresee that it was creating the nervous-system infrastructure for the industries of the twenty-first century.

When the barbarians became bold enough to attack a weakening

Roman Empire, their armies used the very roads Caesar had built for his legions. It was hard not to think of this analogy when terrorists attacked New York in 2001, having used the Internet to perfect part of their plans. This fact alone highlights the need for a much deeper understanding of the nature of networks and their moral neutrality, their effects for both good and evil.

Prelude to Connectivity

Around the time when Paul Baran was drawing up plans for packet switching in Santa Monica, a computer scientist named Robert Taylor had a practical problem to solve in Washington, D.C.

Promoted as head of the Information Processing Techniques office at ARPA in 1966, Taylor saw the need to connect the various computers used by the agency. He walked into the office of Charles Herzfeld, ARPA's director, and pointed out that a great deal of computer development was being duplicated, to the detriment of creativity, efficiency, and the agency's budget.

As told in the excellent book by Kathie Hafner and Matthew Lyon, *Where Wizards Stay Up Late*,[22] Herzfeld gave Taylor the go-ahead to spend a million dollars to come up with a system that would tie the machines together, enabling researchers in widely separated universities to work in cooperation and share expensive resources. The full story is more complicated, since a single meeting rarely leads to that kind of decision. Others were involved in preparing it, and Larry Roberts actually wrote up the ARPA order when he was brought in to run the project.

Baran, in the meantime, had tried to get AT&T to adopt his new communications scheme, but he was unable to move the entrenched corporate culture of the giant firm. As he reported in conversations with Steward Brand,[23]

> I went over to AT&T headquarters—one of many, many times—and there's a group of old greybeards. I start describing how this works. One stops me and says, "Wait a minute, son. Are you trying to tell us that you open the switch up in the middle of the conversation?" I say, "Yes." His eyeballs roll as he looks at his associates and shakes his head.

So Baran sought support elsewhere, at Bell Labs, the Air Force, and the Pentagon. The top brass suggested that the Defense Communications Agency (DCA) should implement the scheme. Baran knew that would never work. Like AT&T, the DCA was run by communications engineers of the old school, who didn't understand digital networks. He decided to wait.

Bob Taylor's network initiative and Paul Baran's packet switching scheme came together when the former hired a hardworking, dynamic, twenty-nine-year-old engineer named Larry Roberts to begin the practical work. Early in 1967 Roberts presented his plan to a meeting of ARPA principal investigators, and the Arpanet was born.

How the Arpanet Worked

Until the creation of the Arpanet, you couldn't get one computer to communicate with another unless they came from the same vendor (typically, IBM, Burroughs, or Digital Equipment Corporation in those days) and were tied to a single process, like accounting or monitoring of industrial operations.

Such computers would be tied together in carefully maintained dedicated telephone lines whose single purpose was to move information from a particular machine to another machine. If something happened to that line, the entire process stopped.

The Arpanet concept introduced a whole new way of thinking about computing. Not only could you exchange information among completely different computing centers using different kinds of machines, but you could write a program in Los Angeles that would run on a computer at Stanford using data supplied by University College in England. And the whole process was extremely reliable because of the way data were exchanged.

Instead of relying on a single circuit, like a telephone line linking point A to point B, packet switching was a diabolical scheme that broke up every message into small units that would be routed by each computer to the "next" one closer to the target. To make this possible, each packet had to contain the address of its destination (say, the computer in England) and a number denoting its number in the sequence, so that it could be reassembled with its neighbors in the total message.

Not everybody loved the concept at first. Big computer vendors like IBM hated it because it would enable their customers to link their machines with those of their competitors. The phone companies hated it because it relied on digital communication, a technique they did not master (they still don't) and on a novel algorithm for moving packets around. Even computer scientists were reluctant at first. Packet switching was very rugged but it did slow down the messages and it required additional processing to route the data around and reassemble them.

Packets also had to be made understandable to many different types of computers coming from different vendors and using different operating systems.

To overcome this added complexity, the ARPA scientists designed special interface machines that switched the packets. Any academic, government, or industrial center joining the network had to agree to acquire this new equipment, learn how to use it, and hook it up to its own machine. The process was not always an easy one to manage, as I would observe at SRI.

About the time when Doug Engelbart's system demonstrated the feasibility and the impact of the first online community, the Advanced Research Projects Agency had just begun to deploy the first network of computers to span a continent, with Baran's revolutionary design that linked together machines of different makes. Computer scientists had to put in new telephone lines to connect their own computers to the spreading network or forget the support of Washington dollars, resign themselves to technical obsolescence, and forever hold their peace. Most of them chose to join.

The objective may have been scientific and economic, but the major impact was psychological: for the first time, a door had been forced open in the bureaucracy and a new type of community had started growing. It was the community that Doug had sensed was going to come, but he was unprepared for its sudden consequences. ARPA decided to renew the financing of SRI-ARC if (and only if) Doug agreed to modernize his programs, change his computer, and hook it up to the Arpanet, whose protocols it would have to obey. Since many of the special functions of NLS could not be moved to the new computer, the project suddenly lost in power what it gained in geographic coverage, by virtue of its access to the vast network the Pentagon was building.

By the time I joined SRI, for instance, it was no longer possible to use the graphics capabilities the team had developed on its earlier machine or to do video mixing with text. "Ah, if only you'd been here two years ago!" my colleagues told me when I raised such issues. Just try to do video mixing with text on the web today, and you will realize that some technologies move *backward,* not forward! Similarly, I remember fondly the late 1970s, when I could travel around Europe with a terminal and connect to the same network with a single account name and password.

ARPA officials never knew all the details of what happened after they pushed SRI-ARC onto the Arpanet: they weren't there to see it, and they were only interested in technological results anyway. The social side of things was of no lasting concern to them. But I was living with the project and saw the most important part of the ARPA experiment from the inside—the human part. Again, it was easy to see that any information network was a social system.

Knowledge Workers of the World, Link Up!

Lenin had told his Soviet cadres to electrify the countryside. The Pentagon decided to computerize it. Its research into miniaturization had already made pocket calculators possible, as well as radio terminals that could communicate wirelessly with computers thousands of miles away. Its orders would now spur the spread of the Wired Nation. But the folks at SRI were torn once again by their identity crisis: why should they take instructions from an organization they regarded as the "Masters of War," when as individuals they were fighting the draft in California, demonstrating against the war in Vietnam, and seriously thinking of moving the ARC computer to the woods of Mendocino to start a programmers' commune?

The confrontation became obvious one afternoon when the group, riddled by conflict, wheeled all the terminals into the corners and spread a carpet in the middle of the main room. The time had come for a real brainstorm. The programmers, in their blue jeans and colored shirts, took off their sandals and sat in a circle.

By the time I joined the project a year or so later, the story had become legendary. It had entered the folklore of SRI, a tale retold to every new recruit, because it exemplified the contradictions under which the group labored.

Once the staff members had assembled, someone stated the issues and invited comments. In the spirit of the time, a bottle of wine and a few joints had been produced; and a serious encounter session had begun when the stairway door opened without warning. Who should walk in but the director of SRI himself, in his gray suit and striped tie, followed by several high-ranking officers from the Pentagon. They were on an official site visit, checking the expenditures of public monies under their jurisdiction.

"And here is our project for the augmentation of human intellect," the director is said to have told his distinguished guests, without even looking. Then he looked, and saw, and smelled, and realized what the unmistakable odor was. He made up some sort of excuse and left in a hurry. Doug's project had just acquired one more crisis.

We must say this about Engelbart: he was able to rise above all those problems. He let go a few troublemakers and replaced them with new people he could better indoctrinate with his vision. He never made clear what the vision was, but he always managed to mesmerize his listeners. I thought I understood the vision: Doug, in his prophetic "right brain" way, saw the day coming when computers no longer would be used by just a few lucky scientists, but would be available to a great many people. He foresaw entire communities working through them.[24, 25]

Doug's genius had grasped the effect of computers as early as 1948. I came to a similar realization much later. It was the suspension of time and space that excited me, and the simple precision of the software as a tool to navigate uncharted waters, new data structures. The NLS system, on a good day, came close to that.

I never had a chance to explain my feelings to Dr. Engelbart. As project leader, he did not have time to listen to anyone very long; he was spending too much time fighting red tape and bureaucracy. Besides, he already had gone on to a new adventure: he was reading a book by management expert Peter Drucker, who enthused about "knowledge workers" as the new international elite. Engelbart embraced the concept and made knowledge workers his new heroes, his marketplace.

Knowledge workers equipped with NLS terminals and thus endowed with think-enhancers would be able to solve, at last, humanity's most difficult problems. At General Motors, the knowledge workers would use NLS to enter their memos and their databases. At American Airlines, knowledge workers would type in their inventory lists or flight schedules and respond to requests for part numbers. At GE and at Corning Glass—in fact, throughout corporate America—NLS would enable specialists to keep track of markets and accounts. Soon every manager would be surrounded by a small battalion of information experts, each trained in the infinite subtleties of NLS: the system, of course, would be much too complex for the managers themselves to use. Beyond all that, communities of researchers, planners, and futurists would tackle the crises of the day—energy and the environment, and even world peace.

From now on, Doug announced, we were doing office automation. And that was official.

The ARC workers took the big news in stride. If Uncle Sam wasn't supporting intellect enhancement anymore, but had big bucks allocated for office automation, then office automation was obviously what we should be doing. So another big meeting was held (minus the evil weed this time), many beautiful feelings were "shared," and we all embraced, moved by emotions and too much wine and too many dreams of a future of peace and harmony. Secretaries who wore no bras wept openly on the shoulders of systems programmers wearing shirts with pretty pink embroidered butterflies (I owned one of those). It was a touching scene of innocence and scientific purity, one that Doug surveyed with benign paternalism, before closing the door to his corner office and privately reviewing old files and journal entries for inspiration.

When the staff came to the office the next day, having had some time to think about the future of the project, there were a few people who didn't feel like crying or dancing for joy any more, and I was one of them. We didn't think that NLS was ready for office automation. We didn't even believe that "knowledge workers," assuming such creatures even existed, were ready for NLS either.

For one thing, we had built an extraordinarily complex system; hundreds of commands had to be memorized by any programmer trying to use it. Mastering them took years, not months. New ones were invented all the time. Far from trying to simplify his system, Doug had encouraged just the reverse: a programmer's productivity was measured by the number of commands he added to the language, which eventually would have to contain more words than natural English. In the mind of its inventor, after all, NLS was destined to replace ambiguous and unwieldy English forms with new constructions that would leave no room for error. In English, you can say sloppy things like, "You wouldn't recognize little Johnny, he has grown another foot."

Doug could not tolerate such ambiguities. One day, as he was explaining this to me, I glimpsed what was happening. An engineer who was deeply unsatisfied with the common language was looking for machines that would permit him to create his own. He had found an ideal environment in office automation, a field in which a brilliant scientist like him could help determine the flow of information among human beings.

Since information is control, those same engineers who shaped information structures in the office of the future would also subtly control office workers, and through them, the whole corporate edifice. This would be done in the name of productivity and cost-effectiveness—two notions easy to measure in an industrial setting, but largely undefined in the office world—and even more unclear when it came to "knowledge workers."

The staff were skeptical of their system's ability to make a contribution to the office world, no matter what the motivations of its designer might be. We could not even run a thirty-member project. The flower children, who were spontaneous and genuine enough to reject the dark and troubled world of their parents, were now simply building another dark and troubled world of their own. One woman who joined the project came to her interview dressed in a casual blouse and skirt, and after looking around at the Levis and the beads and leather jackets, confided to me in mock apprehension, "How could I join this project? I don't have anything to wear!"

We had three typewriters, all in poor condition. The phone was answered irregularly, messages were frequently lost, and it was impossible to get copies made. Yet we considered ourselves a model to communities to come. Why should we conform to the reactionary concepts of ordinary business, which was on the way out anyway? Our advanced computer theories would throw the old office world into the trash cans of history.

Here again, Doug had foreseen the cultural difficulties in which his project would become embroiled. In his July 1970 report to NASA he had already written, with prescient wisdom:

> NLS and our other augmentation systems are part of an exceedingly complex total system that includes all of our set procedures for doing things, our management methods, our goals, and our strategies and priorities for doing research. This total system is the tangible part of the "augmentation culture"; the intangible part consists of personalities, emotional interactions, intellectual orientations, etc.

Observing that ARC was "a tiny and incomplete culture with a brief history" and that "like most cultures, it is incompletely understood by the people inside it," Engelbart's report stated that

> the long-term side effects of any innovation are unpredictable, and occasionally such side effects give rise to problems.

BOOTSTRAP Institute

Figure 8: The Author at the SRI-ARC Computer Lab, 1971

As it tried to cope with the world of office automation, SRI-ARC entered another phase of reorganization and social experimentation. The secretaries decided they shouldn't be stuck with menial tasks like typing and answering the phones, too indicative of the lowly position of women in business. The male programmers in the name of the Knowledge Revolution agreed, and left their terminals to rush to any phone that rang. The result was chaos on two fronts, the phone lines hopelessly confused and the programs tangled beyond repair. The project became less and less productive.

Identity Crises

Doug Engelbart may have found it hard to solve his project's identity crisis but ARPA in Washington had the same problem: they had trouble justifying to Congress their continued involvement in computer research. At an annual budget review, a Defense subcommittee called Larry Roberts on the carpet to ask how long this "experiment" was supposed to last, and why the hell the Pentagon was entering the field of office automation when IBM and Xerox, for starters, each had multi-million-dollar projects to do the same thing.

ARPA's computer budget was cut. Overnight, it became anathema to mention office automation around the Arpanet.

Major projects were reabsorbed into Defense-related missions, and the SRI group was left high and dry. The researchers still made big speeches about knowledge workers, sounding more and more like a

Figure 9: The Author (on the right) with Doug Engelbart and Elizabeth "Jake" Feinler, 2000

pack of wild dogs howling at the moon. Washington was no longer listening: Doug was in danger of losing his financial support. Simple tension grew into open conflict, and the project became stranger still.

Engelbart had to do something to recapture his weakening leadership. A new fad was sweeping California, Silicon Valley no exception. It was a movement that superseded polarity therapy, Rolfing, Esalen, and even psychic healing. Doug embraced it as a timely solution to his problems. It was called "Erhard Seminar Training," or EST.

Those who wandered from group to group looking for cosmic truth, the veterans of the great drug explosion of 1968, the seekers who were forever finding and forever joining new movements, were finally engulfed by this absolute end to the search: they had seen the great light at last. Doug was among them. He sent his whole staff to take EST.

In my opinion, there was a good match between Engelbart's vision at that time and the philosophy of John Paul Rosenberg, alias Werner Erhard. Doug's work had been devoted to giving people a machine through which they could communicate and a language that would tolerate no ambiguity. Erhard was achieving control of the thousands who attended his seminars by a purely psychological technique and a single focus: his own word.

Doug was all intuition, and he mistrusted dialogue, but he was a master at controlling the channels of information. Erhard funneled everything through a single channel, and he was extremely good at words. This former salesman of the "positive thinking" school had carefully studied the movements of the New Age consciousness, from the hot baths of Esalen to the group exercises of Mind Dynamics. He packaged it all into an intense experience, even sprinkling in a little Zen and Scientology. It became an overnight success.

Taking two or three hundred people at a time—because it is much easier to control a group that size than a smaller one—the EST trainers would lock the meeting doors, physically cutting off access to the telephone, the candy machine, and the bathrooms, and denying their students the materialistic pleasures they had always counted on. The trainers then proceeded to break down their defenses, using standard techniques that amazed and disoriented the uninitiated. The EST trainers capitalized on their sense of failure. They would force people to recall the most humiliating experience of their lives and "share it" with the group. They would dismember them mentally, then put them together again like broken dolls.

The stories about Erhard's sessions had interesting effects on the ARC group, over a month or so. First came the gentle suggestion, the subtle pressure for everyone to register for the seminars.

Nothing ever got done at SRI-ARC in those days without lengthy arguments. Doug himself rarely attended the meetings. They were run by his assistants sitting cross-legged on the floor, eating potato chips and drinking Coke. This was a way of expressing their identity, making it clear to the whole world that though they might be taking millions of dollars from the Defense Department to run their research, it would be wrong to confuse them with the lackeys of the Masters of War who worked on laser-guided bombs and who were recognizable by their ties, shiny shoes, and normal mealtime behavior. The ARC project was different, and they were going to prove it by enrolling in EST, a process about which they knew very little. The great mystery loomed above them, and it was their opportunity to solve their personal and group conflicts.

A dozen staff members caved in right away. They were ready for it. They had been calling for it silently. Their whole being welcomed it, their minds gave no resistance. They were professional people, for the most part, who accepted the blame for the current failure of the augmentation project. Since Doug could not be wrong, they themselves must have been unworthy of his great plan, they must have failed in the great mission he had given them. It never occurred to them that the project had become mired in the swamps of its own confusion, that it was losing touch with the mainstream of computer research.

These believers dived into EST like a group of travelers long lost in the desert, seeking salvation in the waters of some refreshing and mysterious river. When they came back to the office on Monday, it was hard to recognize them. What they had undergone was not just a simple attitude change, it was a transfiguration. They glowed, they floated, they hovered, they levitated. Nothing got done that day. They bathed for hours in the ebullience of their new spirit, and the others listened to their stories, which were couched in a peculiar new language that clearly eluded and excluded the rest of our group.

As soon as you began talking business with one member of the first wave of EST experiencers, you would be interrupted by another member who rushed over with some "space" to "share." The two of them would start reminiscing about the deep, wonderful secrets of the past weekend. My reaction was to go away on tiptoe, like a profane tourist who, looking for the exit on the side of the cathedral, has stumbled upon a meeting of archbishops discussing the transubstantiation of the soul in Latin.

Other mysteries were implied by the proselytizers. They now knew how to be permanently happy. They had clothed themselves in an otherworldly glow. They would never be late for meetings again; they were going to be healthy forever. They had achieved a superhuman state and had sneezed their last sneezes.

The First Wave put so much pressure on the group that a second splinter began to come loose from ARC. To me, watching from the outside, EST had become like a wet blanket of conformity thrown over a nice team of gifted individuals trying to live out Doug's genial dream. The augmentation idea had now been restructured to enclose even bigger and more inaccessible goals: universal happiness, permanently clear thought. Scientologists spend years and many thousands of dollars trying to get "clear"; EST covered the same ground in a few days for just two hundred and fifty bucks.

For those who had not yet taken EST (you said, "I'm taking EST" as you would say, "I'm taking five hundred milligrams of penicillin"), the pressure became unbelievable. For a while, it was constant bombardment from group members who had gone through the Mysteries and now had a stake in the outcome. The whole world must go through EST to be saved, it was explained to me, or at least the whole ARC project. If just a few black sheep like me refused to go, they said, that was enough to ruin the whole effort.

There was another reason as well, a hidden one. They had gone through the humiliation, the stripping, the public excision of their souls, the animectomy. They had been given invisible badges of genius in return for their debasement. Now the others had to do it, too, or suffer social rejection. That was the underlying rule of any initiation, any group secret. In spite of their participation in the country's leading experiment in networking, the staff members at SRI were as vulnerable to this psychological pressure as any other group.

My status in the team was peculiar. I was from industry, like the more senior managers at ARC, but I was younger, and was known for my computer work among the peers of the hip programmers. EST wasn't for me, although it was perfectly fine with me if they went to the seminars. I wasn't perfect, I admitted, I didn't think clearly at all times. I caught colds and was occasionally late for meetings, but I had my own standards and would work on my inadequacies without Erhard, thank you. So far as the rest of the project was concerned, the believers expected no organized resistance and they found none.

Instead, they found individuals who could stand on their own two feet and simply told them to get lost. This was a revelation to me. I discovered the strength and the resilience of some team members whose real spirit I had never suspected, and I was even grateful to EST for having revealed them. They were quite a group.

There was Cliff, who ran the network and was our main professional contact with the ARPA community, our gateway to Washington. He had a thorough knowledge of the protocols that kept the computer accessible, and he would have made an ideal recruit for EST. Unfortunately, he had already chosen his own spiritual discipline. He was studying the mystics and had no intention, he said, of spending hours listening to a bunch of pathetic materialists looking for two-hundred-and-fifty-dollar shortcuts to enlightenment (the fee went up shortly after).

Then there was Guru, a very talented programmer, a big man who had often displayed a tendency to insubordination. He resisted EST, and this came as a big surprise because he had long hair, wore the right kinds of clothes, kept a guitar in his office, and talked dirty. True, they could fire him, but he was the only one in the group who knew the computer's operating system. If the software was not maintained regularly, it would collapse and NLS would crash. A delegation was sent into Guru's office to put gentle pressure on him. After all, SRI was paying for half of the fee for EST, and if he was short of cash, maybe even that could be handled.

Guru told them that he had a nice curvaceous girlfriend in the hills and had "better things to do on Saturday than listening to a bunch of hustlers." If they liked to pee in their pants in some sweaty meeting room in front of two hundred idiots, however, he wasn't going to stop them. This particular style of domination didn't turn him on, he said. "I've tried it and I didn't like it." He waved aside his shoulder-length black hair and returned to his computer terminal.

Charles was the next to refuse the proffered cup of EST bitterness and joy. He had long hair, too, in the hip fashion of the day, and he was the only remaining programmer from the team who had written the current version of the NLS program, so it was unthinkable to get rid of him. He came to SRI on his bike whenever he pleased, got the work done, and went home. On weekends he made furniture. His real goal in life was to become a woodcrafter. He smiled when invited to EST, and said in his soft-spoken way that he had a Louis XV chair to finish before he would consider anything like that.

Then there was Mil. She had the office next to mine. She was a strong, cheerful person, usually dressed in purple, and I knew why she picked that particular color and why she wore certain stones on certain days, because I had some knowledge of astrological lore. She had confided to me where the hexagrams were on the floor of her office, and her shelves were well stocked with the occult books of Manly Hall and Dion Fortune. When the computer crashed or when Doug got into one of his detestable moods, I went into Mil's office; she gave me some herb tea, and she told me her thoughts on the tarot and the Holy Grail. She even had a scheme to run obscure cabalistic calculations on the SRI machine someday.

Mil had studied the esoteric traditions for many years, and she regarded that "kid Erhard" as one of those Werner-come-latelys who had missed the whole point about initiation.

The apostles of EST realized they were wasting their time trying to recruit us. The project conflicts deepened as time went on. The glow of the EST experience gradually wore off. ARPA threatened to cut off our funding again. People started coming to work late. And then one day, right in the middle of a staff meeting, Doug sneezed.

The Engelbart Legacy

The EST experiment had been useful, even if the results were not those Engelbart had hoped for. It had not helped us to augment human intellect—a task best left to future generations of gene therapists—but it had revealed some hidden faults in the project. It did some good for those who took the seminars, because the idea of taking responsibility for their lives was probably something they had never encountered. It also did some good for those people who declined to attend, because it gave them the special strength of knowing they could stand up, alone if necessary, and preserve their own standards against enormous group pressure to conform—the type of pressure the Solid State Society is going to place on all of us very soon.

Seen from that perspective, the project was a portent of things to come, a prototype of a world where advanced computer technology combines with advanced mind-control science to enforce desired behaviors. Perhaps we cannot avoid going through such a phase before we discover a higher level of human freedom. For me, the lesson was the revelation that a few people could successfully insist on preserving their personal freedom against the machine.

The results were predictable. Armed with their new strength, those who had not taken EST began to look around. I asked myself what I was doing there. I resigned to head up two projects Paul Baran had initiated at the Institute for the Future. I had spent less than a year in the SRI project. Cliff, Guru, Mil, and John also resigned. What was left of ARC floated on for a little while, and then SRI managers succeeded in selling the project to a company that was looking for a document-preparation system. It had truly shrunk to that in the eyes of the business world, no more than a text-editor similar to those you will now find in any large office. Once again, computer technology had devoured its own children.

It's hard to give full credit to Engelbart's vision, for the human equation has obscured the technology, and the new computers have now made obsolete much of the work of his group. Engelbart's project must be recognized for inventing the mouse, hypertext, a primitive form of windowing, and the first example of groupware, the collective journal of an online community.

The project also made an indelible impact on the Arpanet and the young Internet, because we were responsible for building and running the world's first network information center, which we lovingly called the "NIC."

Under Doug's patronage, one of my most interesting tasks was to design and program the database for the NIC, the first list of sites on the ARPA network. It allowed the early users to scan all the services contributed by the various organizations connected to the network. It is from this rudimentary database that the INTERNIC, the central repository of resources on the Net, evolved into the form it has taken today.

It was a joy to work with the colleagues who assembled the information for the NIC database: such people as Elizabeth "Jake" Feinler, who went on to play a key role in network standards and originated the expression "dot-com" for websites dedicated to electronic commerce; Charles Irby and Harvey Lehtman, who would put their mark on many networking projects after SRI; and numerous others who leveraged their experience with the project into brilliant computer, research, or business careers.

Another important effect of ARC has been indirect—the inspiration of the many designers and programmers who visited our lab over the years and have gone on to other jobs, gratefully recognizing many years later what they had learned from Doug's vision. Silicon Valley is full of such tales, and somewhere near Santa Clara, Doug Engelbart goes on thinking of a time when computers will eliminate war, poverty, and the ambiguities of human language.

In the world of programmers, however, there is something dangerous and dark about building a human communication machine. Many human beings cannot stand direct communication, and some of the brightest technical minds avoid normal dialogue. The little terror that lurks in a dark corner somewhere in all of us, of the "Other Person," can be alleviated by a machine that does our communicating for us; witness the ambiguities of e-mail and chat rooms on the web today. The NLS system, in part, could be seen as a response to that terror. It was a machine we could put between ourselves and the world of others. "Can I reach you this way?" You would push a few buttons, watch the little lights. "Or can I reach out that way? How can I make sure that your words will not harm me?"

The real purpose of NLS may have been to build a screen for its designers, to shield them from a terrible, frightening world. The system would remember everything about its users; it would keep an accurate profile of them, it would keep a searchable record of everything they typed, it would offer you a path to the thoughts of others. The knowledge of that path was power; and for that reason NLS had to be very complicated, it had to be obscure to the casual user.

Is there a similar motivation behind much of the Internet today? Behind its congenial facade, is the social need answered by the network actually one of separation, of segregation, of increasing control?

I believe the truth isn't so gloomy. When all is said and done, when machines have tried in their simplistic ways to solve the human misunderstandings, people will go back to older ways of dealing with one another's thoughts. They won't trust their transactions to the inner workings of a system controlled by armies of programmers. Instead, they will set aside the shields of their software, and in my own dream, they will raise their heads above the curved plastic of their computer displays and patiently learn again to look one another straight in the eye.

Part Two

Making the Planet a Better Place

The consequences are before our eyes: an explosive growth in activity, wealth and power, with technologists so taken by the excitement of their own game that they do not find a moment to meditate on the measure and opportunity of their actions, and on the fragile metaphysical premises that buttress all their buildings.

—C. Marchetti, "On Progress and Providence,"
IIASA (Laxenburg, Austria), September 1977

4

The Birth of a New Culture

The Internet seemed to come out of nowhere in the mid-1990s. In reality it was derived from the Arpanet, which had served as its prototype twenty years earlier. Even today the web is still largely based on the Arpanet architecture defined by Paul Baran, Larry Roberts, Bob Kahn, Bob Taylor, J. C. R. Licklider, Leonard Kleinrock, and a few other pioneers. The first network grew organically from their designs, thanks to a rich pool of ideas fed by some two dozen research teams, and herculean efforts by the engineers at Bolt, Beranek and Neuman (BBN) who built the hardware.

The teams had been working in relative independence, assembling various parts of programs that coexisted on the network. And one day in 1972 they were ready for a public demonstration. To me as a young programmer the event was a revelation, the awareness that an important page was turning in the history of computer technology. The unveiling of the Arpanet would change more than the structure of the technical community. It would soon redefine industrial and social communications.

Chart 2 summarizes the main events that led to this momentous demonstration. Two things are remarkable on that chart: first, the absence of the major computer and information companies of the day such as IBM, Control Data, or Xerox; second, the overwhelming role of ARPA's project monitors and program managers in guiding the development of the technology.

Washington, D.C., October 24, 1972

The first public demonstration of the Arpanet took place at the Hilton Hotel in downtown Washington on October 24–26, 1972, at the International conference on computer communications. The event has remained as a watershed date, somewhat like October 4, 1957, for the launch of Sputnik and the beginning of the space age.

Until that demonstration in 1972, the various development projects funded by ARPA around the country had been pursuing their own research. They did share some computer resources and data (that was, after all, the very definition of the network), but they did not rely extensively on each other's results for their own progress. Even Doug Engelbart's NLS system had been designed as a local project; its connection to the Arpanet was an afterthought, and many remarkable features had been lost in the process. No wonder the assembled scientists (who had been working since late 1971 to prepare the demonstation) were shocked when nineteen teams began demonstrating their systems under the same roof.

The Hilton had given Larry Roberts and the ARPA group less than forty-eight hours to install the network connection. The Boston firm of Bolt, Beranek and Newman (BBN), which implemented the hardware and software interface, managed to bring up the site overnight. By the time those of us who worked on the research teams arrived, the large room was still crawling with cables and connectors but everything worked. That, in itself, seemed miraculous.

Doug Engelbart had sent a small group of us to the event. I was sharing a room at the Hilton with Paul Rech, a colleague trained in economics and a former operations research expert at Shell. As we surveyed the scene and compared notes, it dawned on us that the demonstration would change the architecture of communications, not only in terms of hardware connections, greater bandwidth and reliability, and lower costs (the initial ARPA objectives) but in terms of human interaction and the structure of large enterprises. Many of our colleagues had a similar catharsis that day.

I have saved the brochure we issued to all the attendees on behalf of the newly formed Network Information Center for which SRI had

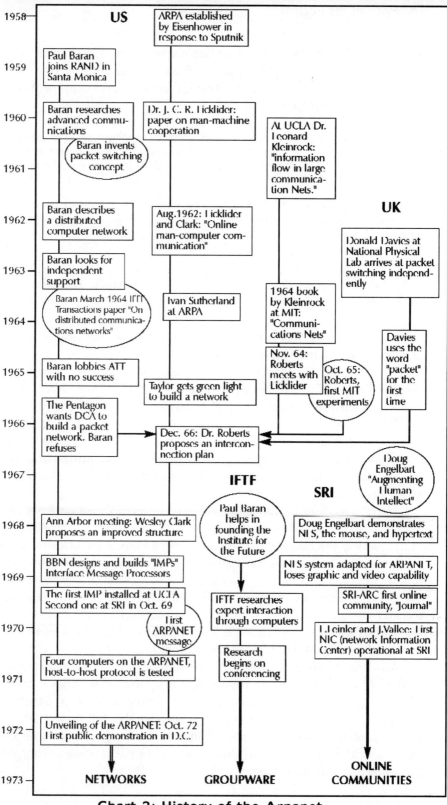

Chart 2: History of the Arpanet

primary responsibility.[26] It is a quaint sixty-two-page document with yellow covers, plastic binding, and a short introduction by Bob Metcalfe, inventor of the Ethernet protocol, then a researcher at the Xerox Palo Alto Research Center:

> We intend that the following scenarios be used by individuals to browse the ARPA computer network in its current stage of development and thereby to introduce themselves to some possibilities in computer communication.

As a computer scientist experienced with the vagaries of such public demonstrations, Metcalfe, who edited the document, was careful to add:

> The scenarios are by no means complete or perfect. We have tried to make them accurate, but are certain that they contain errors. The scenarios are, therefore, only one kind of tool for experiencing computer communication.

In other words, this is research, nobody is sure it's going to work, and don't complain if there are still bugs.

The range of programs that were demonstrated was remarkable, even by the standards of later networks. There were four systems that conversed with their users in English, namely, "Doctor," which simulated an interview session with a Rogerian psychotherapist; "Scholar," for testing students' knowledge of topics such as the geography of South America; "Parry," a Stanford AI program simulating paranoia; and "Timmy," implemented at UCLA as an experiment in natural language processing.

There were four online games: two versions of Chess (at BBN and MIT), a program called "Life" from BBN, and "Jotto" from MIT. "Life" was a mathematical simulation of a colony of grid-dwelling organisms. In "Jotto" you had to discover a secret word picked by your opponent.

Other programs were purely technical tools. They included the file transfer protocol, designed under the leadership of Abhay Bhushan from MIT, various programming languages from Harvard and UCLA, and Mathlab's symbolic algebraic manipulator.

All these programs could be accessed via the Arpanet terminals at the Hilton, but each ran on a single machine. Thus BBN's chess-playing program ran in Boston, while Parry ran in a Stanford computer. The only programs that illustrated the full power of the fledgling network were the network information center we had implemented at SRI-ARC, and the UCLA display of packet flows from the Network Measurement Center.

The Soul of a New Network

The Arpanet was designed to grow organically. As such, it had no center. But if you had to point to an indispensable repository of data that was essential to run the whole net, it would be the Network Information Center.

The Network Information Center was a database system, so the task of building the first implementation went to Elizabeth "Jake" Feinler, a library science specialist by training, and to me as the developer of several generalized database systems at Northwestern and Stanford. The task involved as much coordination with other projects as it did new programming on the SRI machine: the NIC had to provide an index of computers, terminals, and programs for any Arpanet site; it had to drill down to site-specific data with information about that installation's software, hardware, and service configuration, as well as staff names and phone numbers—truly a foreshadowing of what the web directories offer today.

The ARC project had volunteered to implement the NIC as a

Figure 10: Four Pioneers of Computer Networking
From left: Vint Cerf, Bob Kahn, Len Kleinrock, Larry Roberts. Reunited on September 2, 1999, at UCLA. In the background is the first Arpanet IMP (Interface Message Processor) installed thirty years earlier.

Photo courtesy of Dr. Len Kleinrock

demonstration of what Engelbart's NLS could do. Indeed, in building it, I took full advantage of the elaborate structure of NLS data, so that users could simply request to "show" the details of any branch as they went through the tree representing the sites.[27] The result was a system with so few commands that it ran somewhat counter to Doug Engelbart's philosophy, which measured the power of a system by the number of commands it could accommodate. The text structure itself was already so powerful that it enabled me to hide the complexity of the search from users, who came from every site on the Arpanet and had no motivation to learn details of our specific language. The simplicity of the interrogation was a key factor in its success, and the NIC soon became one of the most-used systems during and after the Washington demonstration.

It is interesting to review the scenarios we had prepared for the use of the NIC. They offered users a complete view of all the hardware on Arpanet, all computers and all programs. You could delve into the interests of any particular site, but you could do much more. Using a software "locator" built into the NIC, you could start "browsing" across the net, share files, move them from one site to another. To my knowledge, this marked the first use of the term "browsing" as applied to network information. Yes, 1972 was indeed an interesting year.

While I spent most of my time at the conference hovering around the SRI terminal and explaining the Network Information Center to people who drifted by, I found it fascinating to observe the luminaries of American computer science walking around the room. They were trying to use one another's demonstrations in an unfamiliar setting, and they were failing like undergraduates confronted with their first test in quantum mechanics.

Not only were the operating systems that ran the machines widely different (MIT ran Multics, a predecessor to Unix, while SRI ran Tenex and Stanford had its own system) but even such a simple thing as character echoing could hide unfamiliar pitfalls: some computers waited for the end of a word before sending it back, while others worked like a normal typewriter where each character is typed out as soon as you hit it on the keyboard. As a result you would come across full professors who needed to be retrained before they could type on a simple terminal!

The experience of watching this network of invisible remote computers come alive around us was sobering, often alarming, and occasionally exhilarating in ways that are difficult to recapture, spoiled as we are today in our environment of computing affluence and better standards.

The Washington demonstration was a great success for ARPA managers, a shot in the arm for developers like me, and a chance for principal investigators like Doug Engelbart to hone their research proposals, but it also a lesson in humility. It left everyone with a clear sense of the long road ahead. For the first time, visionary scientists with global connections were confronted with the full scope of the opportunity. Not only did the concept work, but also it suddenly made possible an entire universe of new services. Computers had been successfully linked together in a sustainable, indestructible architecture. The next step was to link the human beings behind the machines. It would start modestly, almost as a game, as a substitute for telex and phone calls among console operators and systems programmers. And it would spawn a new industry.

From Software to Groupware

Electronic mail was not yet a commodity when the first demonstration of the Internet took place in Washington. That very year Ray Tomlinson at BBN had written the first network e-mail program, establishing the use of the @ sign in addresses. It was used by a small circle of technical workers around the net. In July, Larry Roberts had created the first system that could list messages, forward them, and reply; but old habits die hard. Researchers and contractors continued to use conventional methods of communication in their business with ARPA. In fact, there wasn't a single e-mail scenario among the Washington demonstrations in 1972. The network could do algebra and speak in natural English, but one user couldn't conveniently learn how to send a simple message to another.

The turning point came when Larry Roberts broadcast an edict: he would answer any letter within a month, any phone call within a week, and any e-mail within a day. Since he controlled the Arpanet budget, every site manager became an instant convert to computer communication.

For many people, however, e-mail solved only part of the problem. It is nice to be able to send messages around, but that doesn't work very well for groups that need to stay in touch over long periods of time, make decisions about collective resources, keep one another informed about issues being debated, or resolve ongoing crises. Something far more sophisticated was needed.

It was called conferencing.

Doug Engelbart had foreseen the need for conferencing in the design of his NLS, where the project staff made daily contributions to a running journal. But this required mastery of the ARC system, and it only ran on one computer. The need for wider, network-based conferencing systems had been felt by other people, and primitive forms of it were coming into experimental use.

One of the scientists pushing for the development of group communications through computers was none other than Paul Baran himself, who had been so instrumental in getting packet switching accepted.

In March 1973 I met Paul Baran for the first time. I was looking for a job. I had reached a point of utter frustration in my work at SRI. I had been heavily influenced by Engelbart's vision, but I thought the ARC project was bogged down in social issues. I was anxious to apply the lessons from SRI in new ways. Paul Baran, who was a cofounder of the Institute for the Future and had launched two computer projects there, was also looking to move on. He felt entrepreneurial urges after a brilliant research career. He had decided to start a new company. The Institute was looking for someone who could take over the leadership of his two projects and build upon what he had started.

Paul Baran is a quiet inventor and, in view of his many accomplishments, a remarkably humble man. He once cautioned an interviewer who congratulated him on "inventing the Arpanet" that technological development is like building a cathedral. "Over the course of several hundred years, new people come along and each lays down a block on top of the old foundations," he pointed out. "If you are not careful, you can con yourself into believing that you did the most important part. But the reality is that each contribution had to follow onto previous work. Everything is tied to everything else."

That day in 1973 Paul Baran was thinking of laying down a few more blocks. His projects at the Institute for the Future had to do with the creation of innovative software that would go beyond the linking together of computers: it would link people over the Arpanet. Not just programmers and engineers, but experts in scientific disciplines, in economics, in policy management, in the arts and humanities. Paul Baran spoke of "making the planet a better place."

It was a dream that was shared by two other men at the Institute: its genial president Roy Amara (the same man who had played such a discreet but important role while at SRI in getting Engelbart's early ideas funded by the Air Force) and Olaf Helmer, a futurologist who had invented a method for integrating the opinion of a group of experts, the "Delphi technique." The National Science Foundation, who made extensive use of experts in developing its research programs, was interested in improving the process.

Such ideas were typical of the kind of research the Institute for the Future was conducting in the early 1970s. IFTF (as it has become known) was a low-profile think tank, a spinoff of Rand and SRI that specialized in medium-term forecasting for corporations, government agencies, and private corporations. Still in existence in 2003, it has greatly expanded its work in the private sector, looking at new media and the social impact of potential developments in everything from healthcare advances to new economic models.

Located in a magnificent wooden building on a hilltop beyond Stanford, IFTF offered a unique work environment. A friend once compared it to the mood you would find in a monastery. After the hectic days and nights at SRI, where the tensions of the technology were magnified by human conflicts, IFTF was a peaceful haven. Yet the problems before us were serious ones. The freedom to think and create was confronted by daily challenges from colleagues and clients trying to peer into an uncertain future: there was a war in Indochina, Watergate was just around the corner, and America was about to run out of gasoline.

A few weeks went by, and I joined Paul Baran, Olaf Helmer, and the project engineers, Rich Miller and Hubert Lipinski, in what we expected would be the first computer implementation of the Delphi technique. What we discovered precipitated our team into the political turmoil of the nascent Arpanet and launched us into the new industry of social interaction through networks, decades before AOL and the web.

Networks in Transition

Tuesday, May 1, 1973, was my first day at the Institute for the Future, heading up the teleconferencing research, and my first day as a principal investigator on the Arpanet. I had parted company with Doug Engelbart and his group on friendly terms, anticipating that we would share our research results. I hung up a reproduction of the *Unicorn in Captivity* on my wall and organized my files in a large office with Califonia-style windows overlooking a wooden deck and the pine-covered hills on the other side of the linear accelerator and the San Andreas fault. And I started to read the early reports from the project.

The idea of "computer conferencing," I learned, went back all the way to the Berlin Crisis of 1951, in many regards a classic management emergency. All the Western countries had to be consulted by the White House, but there was no time to fly their leaders to a central location. A telephone conference call in seventeen languages would have been an impossible enterprise. The State Department did have Teletype channels to each country, so an attempt was made to splice together the wires to create a crude information network. The result was chaos, with any character immediately duplicated on all the teletypes.

In the end the crisis was handled the old way, through the drudgery of multiple messages and partial phone conversations.

Analyzing the mess in the following weeks, some bright analysts in the Defense community suggested that it should be possible to handle such crises by putting a computer in the middle and using its memory and logic to organize the flow of messages and data around the various sites. Such an idea was easier to propose than to implement. At the time there were no computer networks, no time-sharing systems to enable a machine to service several users simultaneously, and no convenient terminals. There wasn't even any notion of primitive electronic mail.

Bravely ignoring these shortcomings, the Institute for Defense Analyses, a think tank for the Intelligence community, began a research program into what began to be called teleconferencing. In the late 1960s, other groups also started experimenting at the State Department and at the Office of Emergency Preparedness in Washington.

Given a group of people at different locations and a common need for information, computer conferencing is a new medium, a technical tool enabling that group to interact either simultaneously (as in today's primitive chat rooms or instant messaging) or at different times (like a message center, a bulletin board, or an e-mail group list). The link would consist of terminals, a local telephone call to a computer that controlled the process, and a network program that would give users the ability to enter their information into a permanent record of the group conversation.

All we had to do, Paul Baran told us, was build it.

The Birth of a Network Culture

The following July I was back in Washington, this time as a project leader ready to defend a million dollars' worth of research proposals before ARPA and NSF, the National Science Foundation.

The weather was characteristically hot and muggy, traffic impossible. On the same trip I went to the Office of Naval Research and the Department of the Interior, two organizations I thought could use our system once it was perfected.

A big storm blew up in the afternoon, bringing rain in buckets. I welcomed the relief from the heat, but I understood why French diplomats used to consider Washington a hardship post in the early days of the Republic, when Paris sent off people in disgrace as ambassadors to the United States. Washington was nothing but a tropical marsh, with mosquitoes and provincial bureaucrats. It hadn't changed much, except that the swamps had been drained and a few of the mosquitoes had been killed.

Over the weekend Roy Amara and I had a pleasant dinner with teleconferencing pioneer Murray Turoff. But the following Monday we had a depressing visit to ARPA. Research budgets were being squeezed again. Washington was englued in its summer weather and the lingering crisis of Watergate. Nobody appeared in the mood to finance creative research. Our project went on, as did the early design for the Internet (Vint Cerf and Bob Kahn spent 1973 on the problem), but the momentum of enthusiastic development seemed to be slowing down.

There was a malaise around the net, and it wasn't just a matter of technology. It was turning into a social system with its own cultures and subcultures, a foretaste of what the world would later find on the Internet.

I first learned about the hidden subcultures of the Arpanet when a former colleague from SRI drove across town for a friendly lunch with me. She told me there was an underground computer group that operated in the shadows of the network. It dealt in esoterica, magic, advanced software, cryptography, and occasionally exchanges of drugs for money, using the network facilities. The members were among the hackers who spent their nights implementing various features nobody had ever intended the network to possess, programs with the capability to project themselves from one site to another across the country, ancestors of today's viruses and worms. The group called itself the Midnight Irregulars. Among other things, it had just invented e-commerce.

When I flew back to Washington in October 1973, professer Licklider had replaced Larry Roberts as head of the information processing technology office that ran Arpanet. This time I visited the Education Institute, the Institute of Mental Health, and the Department of Labor as an evangelist for computer conferencing. My talk fell on deaf ears. Those institutions were still in the Dark Ages in terms of computer technology and had little interest in joining any network. There were other concerns in Washington: a murderous war was taking place in the Middle East, Nixon was contemplating an American military expedition to Suez, and the government was sinking into fresh scandals. I did meet with Larry Roberts in his new role as president of Telenet to brief him on the progress we'd made with our system, which we now called Forum, and about potential business applications of computer conferencing.

The following month I was back in Washington to brief our project managers at ARPA, where I met Connie McLindon, one of the few women at the top of the computer community. An astute administrator with a keen sense of humor, she took the time to explain to me the obscure meanders of the bureaucracy and introduced me to her boss, the director of ARPA, Dr. Steve Lukasik. It was Lukasik who encouraged me to develop a full-scale program for network conferencing.

Not everyone was ready for the new forms of communication, however. This was made clear to me in December of 1973 when I flew to France, hoping to spread the word about our research and to recruit international participants for future trials of conferencing.

A friend in Paris had introduced me to people from *France Catholique*, an influential magazine. We met in a beautiful apartment near Montparnasse. They were philosophers who knew nothing of technology and were in awe of computers. Focusing on the fact that the project came from the Institute for the Future, they politely asked me to define "at what level of reality futurology addressed itself." I tried to give them pragmatic answers, which seemed to satisfy them, but it was clear that French society, with the best intentions, was not ready for the culture shock of real networking.

That realization was amplified when French entrepreneur Joël de Rosnay introduced me to the head of a prominent Anglo-French software company that was contemplating adapting Arpanet technology in Europe. The man was a typical French executive, utterly stressed-out, every minute of his day seemingly filled with a thousand indispensable details. He explained to me in a curt and pompous way why network-based communities of the type predicted by Doug Engelbart and our own research in California would never amount to anything.

Cambridge, Massachusetts, January 31, 1974

Giving up on my European explorations, I flew back to the States in time for a meeting of ARPA principal investigators in Massachusetts. The

Figure 11: Members of the Forum Team, IFTF, 1972
From left: Richard Miller, Hubert Lipinski, Robert Johansen, Ann McCown, and Olaf Hefmer.

date was January 31, 1974. I stayed in Cambridge, where a raging wind was turning this Boston suburb into a little corner of hell. It cleaned up the landscape by picking up debris on the road and throwing it against the walls of the motel. It played horrible tunes with the angles and nooks of the awkward building that dared to stand in its path. Periodically it drew a sound resembling thunder out of a stack of metal plates that a construction company had left lying in a vacant lot nearby.

The technologies of computer networking were discussed at the meeting in big words spoken with intellectual definitiveness among ARPA investigators who were building the next level in network architecture. My friend Paul Rech was there on behalf of SRI. Dick Watson, who had left Doug's project at about the same time as I did, now represented Lawrence Livermore Laboratory. The meeting was run by a new manager who was clearly intent on asserting his dominance over the group.

Responding to a suggestion by one of the scientists, he snapped, "Doctor, your idea is unfeasible, undesirable, and even superfluous!" Several investigators within the group privately noted they would have to move on and stop submitting their research to ARPA if they wanted to explore new directions. Paul Baran himself commented to me that he felt much too old for political games.

Discouraged, I went to bed early and slept uneasily for three hours, woke up, and went to the window in time to see a car missing a turn and hitting a pole. Midnight. The wind was howling anew; curiously there weren't any clouds in the sky. The half moon glistened, the stars were pure. On my nightstand was a copy of *The King in Yellow:*[28]

> *Strange is the night where black stars rise,*
> *And strange moons circle through the skies . . .*

The poem seemed to match my mood, with a foreboding of the ugliness to come. In this technocratic world, the future suddenly all too obvious, I felt sad for the generous ideas that would die as people lost sight of their own standards for the sake of money or the illusion of power.

The following month I had another opportunity to present our Forum project at a closed meeting of ARPA Principal Investigators in San Francisco: John McCarthy and Ed Feigenbaum from Stanford, Paul Baran himself, and Keith Uncapher of the University of Southern California were there. I sat next to a pioneer of artificial intelligence named Saul Amarel, whom I hadn't seen since I'd given a lecture at the RCA Labs in Princeton five years before. Doug Engelbart was there, too, but everyone ignored him, his research seen as too esoteric.

They agreed that Doug's insights were brilliant, but people had grown tired of his sermons. His project was beginning to look like a forgotten island, bypassed by the very stream that carried his ideas forward. It would take another twenty years before he would be given

Photo from Author's Collection

Figure 12: The InfoMedia Startup Team in 1980
From left, standing: Lee Womack, Robert Verhey, Sal Suniga;
seated: Richard Miller, Jacques Vallee, Jennifer Lear

due credit and the impact of his early contributions would be fully realized.

The conference was held in a large, square, windowless room in the basement of the Marriott, presided over by Dr. Licklider. The participants were seated around a huge circle of tables, so we had to shout to be heard. The assembly included some pure academics, a few electrical engineers in search of greater engine power who had little interest in software research, and a handful of artificial intelligence researchers whose main concern was to get their computers to play chess. What all of them had in common, besides being middle-aged white men with heavy-duty credentials, was an unquenchable thirst for government funding. On the ARPA side of the table was a manager named Craig Fields who spoke of "transferring technology into the real world."

The Planet System

By the middle of March 1974 we had a full staff of ten people working on the Forum conferencing program at the Institute for the Future. The work was exhilarating, to a great extent because of the quality of the group, with such smart computer scientists as Rich Miller and software genius Hubert Lipinski doing the program development. Dr. Bob Johansen, an astute sociologist with an uncommon viewpoint on futures research, had joined us from Northwestern University (he now runs the Institute as its president). Bob brought new discipline to the endeavor, focusing us even more on the study of computer networking as a true medium for human communication.

The challenge was to bend the primitive network technology of the time to the needs of group communications among specialists in many disciplines—from policy experts to researchers in the humanities or the arts, many of whom had never seen a terminal and had no reason to care about the technology. Another challenge was to convince our colleagues in computer science that network communications were as important as faster calculations or more elegant compilers, and that this field would play a critical role among future electronic media.

The day came when we felt ready for a big test. We launched the world's first conference run entirely by computer. It dealt with the topic of the trade-offs between transportation and communications. It was a timely subject, in view of the gas shortage that affected the entire country during the Nixon presidency. Most of the participants never met face-to-face. The conference was a definite success, and the full transcript was eventually published by Bell Canada, another world first.[29]

Weeks went by, and the lines at the gas pump became shorter, but California—and the whole country—only grew increasingly bizarre under the double impact of bad drugs and even worse political leadership. As I walked downtown San Francisco one day, I bumped into a former coworker from SRI days, drifting on a sidewalk, mumbling incomprehensible words about cosmic messages. He confessed he had spent his last dollar, survived by stealing leftovers from garbage bins, and slept on Skid Row. I invited him over for a decent Chinese lunch. While he devoured it, he explained he had left Doug Engelbart's project to embark on an even more important goal than augmenting the human intellect. His new project was nothing less than a New Gospel. It began when he discovered that he was the reincarnated combination of Saint John and Lord Krishna. "The times are near!" he warned me.

We went out to the littered sidewalk of Eddy Street.

"I've taken over the leadership of the Earth," he said in confidential tones as we picked our way through the trash. Then, on February 13, in a grandiose gesture, he had relinquished control of the planet to Our Lord Jesus Christ.

It was a scary scene, a situation out of some surrealistic film: drifters and pushers, whores and thieves walked by us. In cheap hotels scared seniors huddled around television sets, trying to ignore the dangerous world. We were a long way from the "augmented human intellect" of Doug Engelbart, or the "better planet" of Paul Baran.

Indeed, the times were near. By mid-May Doug Engelbart's project was sinking amid the budget debacle at ARPA. SRI's programmers were moving en masse to the Xerox Palo Alto research center, known as PARC, taking with them the latest mouse interface, the idea of links embedded within text, and Doug's concept of an online community.

Threatened by the same bureaucratic trends, my project left the ARPA environment to expand its research under the sponsorship of the National Science Foundation, where we found a receptive ear among younger, more dynamic managers like Fred "Rick" Weingarten. We decided to launch the next level of conferencing, a streamlined program we called Planet, which we implemented on a commercial network offered by the Tymshare corporation. The drawback was that we had to pay for computer time. The benefit was that we started collecting realistic measurements about real applications, far from the artificial free-for-all of Arpanet research.

On my next trip to Washington in mid-1974, I had the opportunity to meet with Dr. Ruth Davis at the National Bureau of Standards.

She was the most powerful woman in computing at the time, and clearly understood the significance of conferencing. She strongly encouraged us to take our research to a new level by involving scientists and experts from many fields. That same day Roy Amara and I drove out to Reston, Virginia, for a presentation of Planet to the chief geologist and director of the U.S. Geological Survey (USGS) and his top staff. The result would be a series of experiments linking field geologists with their support offices and their headquarters, which gave us the opportunity to quantify for the first time the actual impact of networking on the professional environment of scientific workers.[30]

By 1975 we felt our system was strong enough to be tested in even more advanced communications situations, those involving the minds of participants at the extreme edge of their capabilities. So we launched a privately funded conference on psychic phenomena in which we asked the group members to describe remote targets—a collection of rock samples with special properties, selected for us by a geologist from the USGS.

The conference began, appropriately enough, with a solstice celebration held in June 1975. It was presided over by psychologist Arthur Hastings from his home in Mountain View, conversing over the conferencing network with a group of remarkable people including psychic Ingo Swann, who used a terminal I had installed at his flat in Greenwich Village, writer Richard Bach (the author of the best-seller *Jonathan Livingston Seagull*) typing from Florida, and participants scattered all over North America.[31]

The results of all these trials were so encouraging that we felt the time had come to take network conferencing out of the lab and into the real world. It was time to launch our own software company.

5

Growing the Grapevines

Paul Baran had come up with the observation that "we could build very robust communication networks if we built them like fishnets. But the problem is, how do you get the signal to go through the fishnet from where you are to where you want to go?"

The whole history of the Arpanet, the Internet, and the web is an attempt to answer that question. Every step in the development of the technology has rephrased it, reframed it, and answered it in new ways. The process is still going on. An important step was taken in 1974, when two computer scientists, Vint Cerf and Bob Kahn, published their specifications for an improved way of managing the transmission of data packets across networks.

Kahn and Cerf, who had first met at the University of California at Los Angeles in 1970, formalized the idea while attending a conference in San Francisco in 1973. The problem they needed to solve was how to connect different networks through gateways. They came up with a replacement for the old host-to-host network control protocol and called it the Transmission Control Protocol, or TCP. The task of breaking up messages into packets and handing them to the right program is a very complex one, because the program has to make sure they arrive at their destination and are properly reassembled.

Today every major operating system handles the TCP protocol. It certainly could be improved, made more secure or more robust, but its universality and convenience have made it a worldwide standard. The new protocol was a breakthrough in the building of a reliable medium for worldwide communication, but few people knew about it outside of the academic community. As Vint Cerf observes, "it was not introduced on the Arpanet until 1983. Ten years of very hard work."

In the summer of 1975, after much wrangling over who actually owned the basic code for network processes, ARPA transferred control of the Arpanet to the Defense Communications Agency, with Bolt, Beranek & Newman retaining responsibility for network operations. Thus the military bureaucracy reasserted itself, exactly in the way Paul Baran had tried to avoid many years earlier. The only positive aspect of the move was that computer researchers could now concentrate on more exotic pursuits.

Bob Kahn and Steve Crocker remained at ARPA (whose name, for some obscure bureaucratic reason, was changed to DARPA, only to be changed again a few years later). They managed to lure Vint Cerf from Stanford University to coordinate Internet packet radio research and satellite programs. On October 22, 1977, the team demonstrated a three-network system, linking the Arpanet with a packet radio network and satellite pathways from San Francisco to London and Los Angeles.

In 1978, during a meeting in Marina del Rey, Vint Cerf, Jon Postel, and Danny Cohen came up with a key improvement to TCP. Why not break off the part of the code dealing with routing packets? They called it the Internet protocol, or IP.

From then on, TCP would be charged with the task of breaking off messages into manageable pieces, reassembling them at the other end, detecting and fixing errors. At the same time, IP would do the actual routing across what Paul Baran had called the "fishnet."

Renamed TCP/IP, the scheme proposed by Cerf and Kahn and improved by Postel and Cohen would not be officially adopted by ARPA and the Department of Defense as the core data transmission system on networks until 1983. Although the world at large has seen the advent of the Internet as an instant success, its development actually took five years, followed by five years of implementation and testing, and was marked by many delays and setbacks.

With the ability to link multiple networks of dissimilar computers, and with the robust reliance on TCP/IP, the basic architecture was now in place for the development of the Internet. While such progress was being made in the rarefied atmosphere of computer research, the rest of the world did not seem to pay much attention to the new technology. My colleagues and I had an amusing confirmation of it as we toured the country in search of support for our fledgling start-up.

InfoMedia and the Birth of Conferencing

It is not enough to demonstrate the ability to link machines together. What is important about the birth of networks is the range of new applications they permitted, and the new ways of thinking that came with it. Paul Baran, Doug Engelbart, and a few others had opened the way for the development of what eventually would be called "groupware": information pathways that enabled individuals all over the world to exchange thoughts and data, independent of time and space constraints. Electronic mail, bulletin boards, and online journals were primitive forms of such interaction. Now we felt we could tackle the next job: to construct logical structures that would link together entire intellectual and business communities disseminated around the planet, and to observe their interactions.

We knew we could do it, even in the harsh world of business and industry. Using the Planet system, we already had pushed to its limit the potential of early networks with the help of artists, geologists, philosophers, and businessmen who agreed to serve as experimental groups. Over four years of research, we assembled enough information to distill our techniques in the form of a stable software product. So we wrote a business plan, and in 1976 we launched a company called InfoMedia to commercialize computer conferencing, without listening to negative voices that told us that our business was premature.

We probably should have listened to their advice.

Even in the United States, the networking market was still embryonic. Those industrial companies that had installed a few personal computers were using them for accounting (remember Visicalc on the Apple II?), but they were not thinking of groupware. Even electronic mail was still an esoteric concept in the big wide world of business. "Commercial e-mail was a hard sell in 1983," observes Vint Cerf. "Compuserve, MCI Mail, Telemail, Internet, and e-mail were all separate. They began to be linked via Internet in the late eighties." Electronic mail threatened executive power and was even resisted by the managers of most computing facilities, as a new type of application they did not understand. Most computer center managers came from a mainframe tradition. They were not trained to handle communications; they perceived networking as a "foreign" activity that did not conform to their rules, resisted their ambitions, and escaped from their empire.

While we were digesting this reality, we took comfort in a new phenomenon that had started to engulf Silicon Valley. The venture capital industry, propelled by striking successes in semiconductor development and new electronic architectures, had entered a phase of aggressive growth. It sought new fields in which breakthroughs could be achieved by small teams of scientists equipped with the latest in technology. Stimulated by this renewed availability of capital, entrepreneurs launched independent software companies. They settled in those parts of the Bay Area that electronics factories had not yet claimed.

A new economy was being born in Northern California. It was fueled by innovative financial instruments. Venture funds from Silicon Valley such as the Page Mill Group, Kleiner and Perkins, Mayfield, and Sequoia Partners emulated, then eclipsed venture development on the East Coast, an activity that already boasted a brilliant record.

It was an American officer of French origin, General Doriot, who had made the first foray into U.S. venture capital by financing DEC, the Digital Equipment Corporation, in the Boston area. A wave of pioneering efforts followed, led by men of very different backgrounds. Under the impulse of attorneys such as Fred Adler (Data General), engineers like Kleiner and Perkins (Genentech), businessmen like Arthur Rock and Don Valentine (Apple), and financial analysts such as Ray Williams (Amdahl), new companies started to flourish.

Ray Williams's experience is characteristic. A classic financial executive (he holds a diploma in financial theory from an East Coast university), he had developed the business plan for the 360 computer system at IBM, establishing the framework for one of the most far-reaching technical efforts since the Manhattan project. He worked with a brilliant hardware architect named Gene Amdahl. This ambitious engineer had dreamed of building an even more powerful machine than the biggest IBM 360 "stretch," but IBM refused to finance its development, so he decided to launch his own firm. Ray Williams left with him; over the ensuing years the two men, through a lot of very hard work, built a remarkable computer company, capable of competing head-on with IBM itself.

Having made a bundle after Amdahl went public, Ray Williams became a legendary business angel in the Bay Area. It was with his help and the support of the Institute for the Future that we got some of the early venture capitalists in Silicon Valley to support our idea of using computer conferencing to grow new types of "grapevines" in old organizations.

Our business plan was simple and clean. We would take the computer conferencing technology—complete with an information retrieval engine and instant messaging—into the core of industrial decision-making. Informal networks have always been the real harbor of trust and the spring of action for companies large and small. What seemed powerful to me was the combination of these grapevines, grown explosively to cover the entire world, through casual connections made by invisible wires, with the business resources already in place.

Before action is taken on database information, for example, why can't the decision be available for review in a human conference staffed by experts? When a firm needs to control a project over far-flung regions of the world, why can't the managers stay in touch by having common access to data, budgets, drawings, and each other's thoughts and plans? When a community of industry partners has to coordinate its response to an emergency, why can't it share a single medium where an accurate record is kept, with every member's responsibility clearly identified?

We did have the backing of business angels, but we needed more to launch a real company. After some initial debate, the board of trustees of the Institute for the Future made the generous decision to back our venture. Some of the members were concerned to see part of their research team leave to start a software company, but a heavy-hitter on the board, Uniroyal's chairman, George Vila, stopped the discussion: "If those fellows have the guts to jump into business, it means the work we did has some real application. We ought to be helping them!" They proposed to introduce us to any large company we picked as potential partners in the venture.

We selected four names: IBM, GE, Texas Instruments, and Western Union. Each had special resources that would be extremely useful to us in implementing our plan. While the Arpanet is justly remembered as the prototype for the Internet, several other networks had already been built for time-sharing by the computer industry. The largest one was managed by GE, a splendid company with the most extensive networking experience in the business world; IBM's dominance of computer hardware and software was universally acknowledged, and it had unparalleled credibility with major client firms; Western Union was a global "record carrier" with special telex lines into the offices of every industrial organization; and TI made the chips.

Apart from those four giants, we also had connections to managers at Control Data, a dominant force in high-end computing. We thought each of these companies would have a special reason to listen to a team from Silicon Valley with a new idea in networking.

We were in for a rude lesson in industrial realities.

The First Round

Our search for the first round of investment began in New York, with business tycoon George Vila driving us to IBM headquarters in his little red car. A big and forceful man, Vila drove through Manhattan like a Mexican cabbie, going through red lights while commenting on the varied appeal of passing women, to the horror of the more dignified Roy Amara.

The headquarters of IBM is a big white castle on a grassy green hilltop. Thanks to George Vila's influential connections, we had an appointment very close to the top of the hierarchy, with an executive VP for industrial relations and two of his assistants. I had assumed they would readily understand our product, but I was mistaken. When I plugged in my portable computer terminal under his desk and borrowed his phone to connect myself to the Tymshare network, the executive became visibly confused. And when I started exchanging live messages with a colleague in California, he thought we were playing a joke on him, running a "canned" scenario rather than the real thing.

This was not the kind of thing that had ever been done on an IBM machine.

Once he understood what was happening, the executive asked an assistant whether any work of the sort was under way within IBM. She looked at her notes, and mentioned some electronic mail development in a faraway laboratory, and research on voice systems. I understood her role to be internal intelligence: IBM was so large, and represented such a huge share of computer research, that they needed a special office just to keep track of what went on across their far-flung divisions.

The meeting concluded without a deal. They expressed great interest, however, in "studying" our Planet system, which we took to mean taking apart our source code, so we declined.

The environment for software start-ups in 1976 was not good. I saw that the very next day, when I visited ARPA and USGS and chaired a session on computer communications at the National Bureau of Standards. The deepening economic crisis of the mid-1970s had thrown into the streets a diverse subpopulation of unemployed people and idle young toughs. Broken refrigerators rusted away in dilapidated houses a few blocks from the White House.

Was it reasonable to start a new company when the country was in such bad shape?

A few hours later I was back in New York, which wasn't faring much better. The city was bankrupt; thousands of employees were about to be laid off. At my hotel the gloomy waiters seemed about to collapse from tiredness and boredom. George Vila came back, this time to take us to the headquarters of Western Union in New Jersey, yet another castle on another green hilltop. He walked in as if he owned the place. We met Johnny Johnson there, a former Air Force brigadier general who saw our demonstration, understood exactly what we were talking about, was readily impressed . . . and told us in a tone of apology that they would have to do an extensive market study before giving us an answer. Another impossible six-month delay!

Roy and I then flew to Dallas, where we met with top managers from Texas Instruments. They were dynamic young engineers who walked around in shirtsleeves and did not bother with preliminaries. They looked at the demo, understood every part of it, approved the whole concept, slapped us on the back with great wishes for success, and told us not to count on them for support. They were in the chip business, and could hardly keep up with demand. They loved people like us who invented new software applications. But that wasn't their business.

So we went back to the East Coast for a meeting with GE, which operated an even larger computer network than Tymshare.

I often think back to that meeting at General Electric, because it was surrealistic. Two vice presidents were present with their top staff. Again, the demonstration (which used the competing Tymshare network) went on flawlessly. But it was followed by some sharp arguments among the GE folks. Yes, the concept was futuristic and impressive. Yes, you could build a company on this. But you couldn't use the GE network to do it. Why not? asked one of the VPs. Well, answered the legal expert, it would be illegal. Or rather, it would fall into a gray area, because we were not licensed to send messages around the world, and computer mail was not recognized by postal authorities in other nations. The post offices would feel threatened, the argument went, and access to the whole GE network might get cut off because of us. Tymshare, as a small company, could take that chance, but the much bigger GE could not.

So they concluded the meeting with effusive congratulations for our outstanding job on human interfaces and an offer to come work for them.

Finally, a visit to Control Data headquarters in August of 1976 led to another argument with a legal team that left me angry and puzzled. We began with a demonstration of a new teleconferencing prototype I'd developed with my colleague Bob Beebe for their Cybernet network under an initial contract with Control Data. The demonstration was successful, but it was followed by a sharp confrontation with an in-house attorney who proposed rewriting our contract, taking all the rights to our system for a small lump sum, and leaving us with no royalties.

Even if we had agreed to such absurd conditions (which effectively would have killed our company), our concept of conferencing would never have been adopted inside the company because another branch of the organization felt it "owned" computer communications. That group, called Plato, was part of a pioneering effort funded by Control Data in the field of computer education.

As remarkable as Plato was for education, the system had never been designed for computer conferencing. Yet the arrogance of the Plato systems group at Control Data knew few bounds. They told us they would revolutionize not only education but also business management, reminding me of Engelbart's ambitious plan for reconstructing the intellect and office automation. As for conferencing, they were "already doing it" and needed no new idea from anyone. They only changed their tune when we enumerated all the features they were unable to provide.

What they had was a fairly good "chat" program, certainly not a conferencing system. Yet there was no way to win that battle. At Control Data, in spite of the high standards set by their visionary chairman Bill Norris, the situation was strikingly similar to what we had found at IBM, GE, and Western Union. Each company had a rigid culture, unable to accommodate new ways of doing business, unable even to experiment under the flexible structure a start-up was offering.

If new forms of software were going to evolve, they would have to do it outside the dominant structures. Running on a few dollars from angels and some initial contracts, InfoMedia went into business on the strength of our reputation with a few early converts to computer conferencing and a lot of sleepless nights.

Deals and Ordeals

The new company slowly got off the ground, but we had an initial setback at the Geological Survey when it declined to install our software. The situation was a prototype of many obstacles we would have to face: simply stated, their systems people felt their power was threatened because conferencing would alter social relationships among the user community they served: the network would give local and state geologists straight access to Washington, bypassing regional bureaucracies.

We were running into the conflict between the potential grapevines of network communication and the solid-state hierarchies, well established and solidly entrenched.

The sociological research we had done with Robert Johansen should have warned us about this kind of fear. Many managers in traditional organizations were confused by the new electronic media. That fact had been demonstrated by Doug Englebart's experience, and even earlier by AT&T's refusal to apply Paul Baran's ideas about communications architectures. But geologists were especially rude. At the USGS, a government expert was asked to evaluate the technique for use in mapping and resource studies in Alaska. He reported that computer conferencing in geology would be a waste of time and money.

Fortunately, some of his colleagues were less negative. Under pressure to deploy better communications with their teams in the field, they were even willing to experiment. In the end, our system was heavily used in Alaska, both from the field and to link experts in resource studies. It also saved an international convention of geologists when a Canadian mail strike completely cut off the meeting organizers. There was even one interesting morning in 1975, when a new branch chief was selected through our system, by a live computer conference linking all his colleagues in Washington and Denver.

Two years later the bureaucracy reasserted itself. An internal team decided to build their own system "to save money," so they programmed a monstrosity, loaded with bells and whistles, that only an engineer could master. Everything returned to normal: computer conferencing in the earth sciences was dead. In the process, however, geologists and managers from the oil industry who had observed the use of the system picked up the technique and ran with it. Soon we had contracts with exploration and production engineers of major oil companies who were carrying terminals to the remote offices of their firm, from the frozen reaches of Alaska to drilling platforms in the North Sea, using the conferencing technology to make drilling decisions.

We learned much from the use of conferencing by government geologists. We discovered that effective interaction through a computer tended to reveal the real power structure of a group. That didn't always sit well with the bureaucracy, and it split executives into two opposite groups. Most of them resisted the introduction of networks, but those who wanted to save money and run a tighter ship were able to put the technique to good use. They could pin down a group that wasn't producing; they could get information daily from experts who worked at the other end of the country. On several occasions, geologists from six different countries used our network to set up exchange standards for mineral data.

But the major result of its use was to create, gradually, organizations that did not follow former patterns. *The networks started growing around the hierarchies.* This would be a key factor not only in the success of conferencing, but in the adoption of Internet technology ten years later when it started to come out of the lab. We didn't know what was happening at the time; we didn't have enough experience to see the obvious impact of teleconferencing on human structures.

InfoMedia quickly demonstrated that computer networks could be used for more than computing, that they were a new medium of group communication with exceptional features. But the social and economic features of the technology were still widely misunderstood.

By February 1977 Doug Engelbart was caught in a disintegrating spiral. The management of SRI was no longer supporting him, and turned over his project to another department. They had a champagne party to celebrate the news, even as Doug was holding another long soul-searching meeting of the few remaining faithful "to clarify his future role."

We did not fare much better. By the end of December 1979, Info-Media was broke. Our potential backers kept delaying their decision, speaking of the complexity of their "due diligence" process. I faced two possible choices: either drop the project quietly or draw upon our meager financial reserves to support it myself.

I kept waking up in the middle of the night, tossing and turning, trying to decide between these two courses. It would be a shame to drop the company now, I told my wife, Janine. We had close to two thousand users. Even in the midst of detestable economic conditions, we were within eight thousand dollars of breaking even on a monthly basis! Janine told me she'd agree with whatever decision I made, even if it meant selling our house. We just had to get on the road again and find new clients.

From Planet to Notepad

When NASA's satellite communications video link failed, it was computer conferencing that saved the day. The inexpensive systems we were building might someday connect the whole world, we told their managers. And they would save the considerable investments the space agency was making when trying to support federal efforts for crisis management.

Imagine the following scene: you are in charge of emergency services for a part of Ohio that has just been hit by the worst storms and floods in years. Sixty people have died, hundreds are sick, and thousands homeless. You need five hundred army tents, thirty bulldozers, and two tons of medical supplies. There's a hospital in Dayton that can provide a field emergency facility, but it must be picked up by trucks, and there's a truck company in the next state that can provide transportation but it hasn't returned your phone messages of the last four hours.

Wouldn't you like to be able to tie all these people together in an around-the-clock conference and keep them posted on developments? The communications medium that would make this is called computer conferencing. It is easy to use, and it gives us a way to escape the closed, authoritarian structures computers seem to precipitate.

When you mention teleconferencing to people in Washington, their eyes light up. They think: Live video! Satellites! Big Bucks! Instant Control! We'll be able to find out what's going on in Toledo, Ohio.

Then they realize two things. When you use a network for communication, you don't need television, and you don't need a satellite; all you need is a telephone, an electric plug in the wall and a smallish computer somewhere. It doesn't sound like something you can use to impress Congress, even if it does marvelous things in everyday practice.

The second thing they realize is that maybe they can find out what's going on in Toledo, Ohio, but then the folks from Toledo are going to find out what's going on in Washington, D.C., too, because a network communicates both ways! And if there is one development that administrators want to avoid at all costs, it is having folks in the provinces look over their shoulders.

In my own experience at InfoMedia, one exception to this generally dreary picture emerged at NASA. When we first presented the idea to them, they did not hesitate for a minute to tell us what they thought: come back in ten years. Unlike satellite broadcasting, they said, our network technology was not ready to be used in practical applications.

We did not give up. Among U.S. government agencies, NASA seemed the most logical one to pioneer the use of computer conferencing in real applications. It had many remote sites, a high need to coordinate complex projects, and very savvy technical teams that could readily be trained in using our software. But its bureaucracy was overwhelming. Like most parts of the government, it had gotten used to an established communications infrastructure and had little incentive to change. Furthermore, conferencing by computer was a scary concept. The telephone people at NASA didn't trust or like the computer folks. And vice versa.

It took a crisis to bring them together. The space agency was building a new satellite with ambitious video and voice communications capabilities, and they had trouble getting the project started. There were NASA people in California who encouraged us to send our proposals to the top brass. One day, in the middle of one big satellite presentation to the director of NASA, the television transmission gave out.

Hello! Washington? Can you see us? No, we can't, what's going wrong? Gee, we don't know, are you sure you can't see us?

Boy, I thought. Is that really the space technology of the future? Then the audio itself gave up. Whistles and shrill feedback were all that came from Washington.

We set up our own portable terminal in a corner and started typing. Are you there, London? Yes. Are you there, Chicago? No problem. Slowly the NASA folks came over and looked at our system. It was the only show in town.

"How much does this thing cost?" one of them asked.

I had to repeat the figure several times: less than a phone call to Denver. It was so low they had trouble believing it. They decided maybe they wouldn't wait ten years after all. They would look for the "right project" as an application and give it a try.

The right project came when they launched the Advanced Communications Technology satellite. Not the launch itself; the launch was prepared using the old NASA communication links, but the rocket didn't go up when it was supposed to. NASA faced a terribly important problem: rescheduling the launch party—once, twice, three times.

When you are in charge of launching a satellite, the big event is not the launch itself; it's the party where you will convene members of Congress and their spouses, big money contributors, foreign VIPs, and the heads of major aerospace companies you hope to buttonhole to talk about your next big contract. The party is critical.

NASA got tired of sending telegrams to three hundred dignitaries. They started using our network system as a simple bulletin board, so people could find out the status of the rocket from their office terminal, without calling anybody at NASA. The thing worked so well, the mission director decided to use it to run the project, give frequency allocations, and check on the status of his teams.

It was a very successful computer conference, and it lasted four years; longer than the mission itself.

The successful use of the system at NASA led us to expand the software's capabilities. Rich Miller and a small programming team rebuilt the code under a new name, "Notepad," and launched it with applications to nuclear safety at the Electric Power Research Institute following the Three Mile Island incident, and to numerous other industrial applications. But the company still couldn't break even. The money made on such applications did not cover the cost of new development in an economic environment where the prime interest rate had climbed above 20 percent. InfoMedia could not grow to a level where it could compete with the more established media.

The company had attracted a group of brilliant, dedicated sales and support professionals[32] who brought it to the forefront of the networking world. But we were fighting an impossible battle. By 1982 we had over two thousand regular users of our Notepad system in major companies. Bechtel coordinated eleven projects using the software, including the largest mining construction in the world, the Ok-Tedi operation in New Guinea. Philips Petroleum used it for exploration and production in Europe and Africa. Shell linked together their advanced planners. A dozen other companies had smaller projects, ranging from art exhibits to communications planning. Bell Atlantic, a large communications company where a hippie-guru-turned-management-consultant named Ira Einhorn had introduced us, started experimenting with the medium for executive coordination. But these projects failed to grow into the size that would have allowed the company to flourish. The economics were heavily against us.

When sources of capital became very scarce in the early 1980s, it became obvious that the technology would have to be sold. After a brief, unsuccessful courtship with French giant Thomson-CSF, we were approached by a team led by Renwick Breck, an independent journalist with a vision for a future media company embracing conferencing, news reporting, and emergency management. They bought the system and tried for several years to launch their own concept, a forerunner of America Online. They, too, were unsuccessful, because the market was not yet mature for the adoption of such a medium in the absence of a widely available computer network.

That network would not appear until a decade later, in the early 1990s. It would be called the Internet.

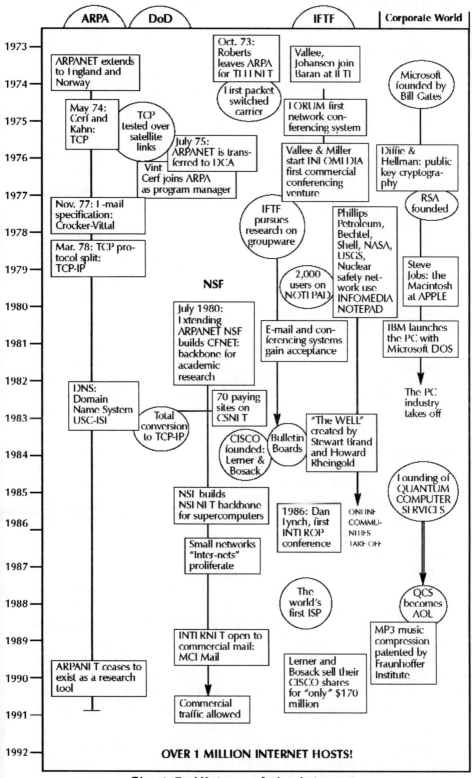

Chart 3: History of the Internet

Transition to Internet

The transition from Arpanet to Internet took place over twenty years. The few books that mention it show this period as a smooth, linear series of dates and events, a standard weakness of most historical accounts. In reality, the transformation was anything but smooth. It involved many diverse interests and institutions. Few of the participants had any notion of what the ensemble would become. They knew they were building a global network, but they thought of it in terms of computer science and improved access to information, not as a novel vehicle for enhanced human interaction.

The big event at the beginning of that twenty-year period was the first public demonstration of the Arpanet at the Washington Hilton in 1972. The big event that concluded the period, marking the end of the transition, was the 1992 release by the Centre d'Études et Recherches Nucléaires (CERN), the big physics research facility in Geneva, Switzerland, of the software that made the World Wide Web possible.

Between these two events there was an enormous amount of work in U.S. and British academic centers and research institutions, much of it funded in parallel by ARPA and the National Science Foundation (NSF).

In contrast with military research agencies like ARPA, or operational groups like DCA, the National Science Foundation is primarily dedicated to the funding of academic research. Its objectives are set by scholars, in a spirit of open cooperation. So when the Defense Communication Agency closed off access to the Arpanet to anyone who didn't have a Pentagon contract, the computer science community in the United States reacted with dismay. Networking had become a way of life for many researchers in fields as diverse as physics and biology, and even in the humanities. They went to NSF with a proposal to build a civilian research network.

First proposed under the name CSNET in 1979, it developed the following year into a scheme to augment the Arpanet with a primitive system based on Telenet and an e-mail service. NSF agreed to support it for two years, after which management would be transferred to UCAR, the University Corporation for Atmospheric Research.

By 1983 the computer science network built by MCI, the University of Michigan, and IBM, became more important than the Arpanet. It had seventy sites online, paying dues for the service. By the time NSF support came to an end in 1986, most computing research facilities in the United States (both at universities and at private sites) were connected, and the service ran at breakeven. Many institutions started building their own networks using TCP/IP and the new technology of "routers," all of which was facilitated by the fact that NSF had also built a powerful backbone.

Although the backbone was initially designed to connect five supercomputer centers (this was called NSFNET), the Foundation offered access to academic institutions with typical charges between $20,000 and $50,000 per year in each geographic area. This quickly led to the creation of a dozen regional networks.

Contrary to what you often read in the press, the Internet was not built entirely with U.S. taxpayers' money or Pentagon funds. The government did pay for the initial research through ARPA and the backbone development through NSF, but academic users had to pay their own way as recently as twenty years ago.

The work to build NSFNET was hard and painstaking, with many setbacks as conflicts had to be resolved among various standards. In the early 1980s there was an uncomfortable argument between the backers of TCP/IP and a consortium of computer companies (primarily IBM, DEC, and HP) pushing their own scheme called OSI for "Open Systems Interconnection." It almost erupted into an all-out standards war. Fortunately, TCP/IP won after huge battles and the Arpanet made a total conversion to it in 1983. Networks using the protocol became known as "internets."

These developments went on without much recognition from the rest of the world, where communications continued to be centrally controlled in the hands of government monopolies. Those networks that prospered in the business world (like Tymnet, Cybernet, or the GE network) made money by giving client companies access to time-sharing resources and databases, largely ignoring message exchange. The word "groupware" itself had not yet appeared.

The state-owned telephone companies in Europe did have their own experiments under way, but they ignored packet-switched networks. At a cost of about $9 billion, France Telecom launched the Minitel, a centralized system that pioneered some services in electronic commerce. To this day, the French are under the delusion that they invented the Internet because of the Minitel. In reality, their technology came many years after Arpanet. Furthermore, it represented the exact opposite of the Internet concept, a closed system with no ability to grow organically. As late as 1996, France Telecom was still trying to prevent the development of the Internet in France. Nonetheless, some original work in France had been done by a brilliant computer scientist named Louis Pouzin.

I had met Louis when he attended the 1972 unveiling of the Arpanet and exchanged research data with him afterwards. He had almost single-handedly built a packet-switched network named Cyclades. The French communications bureaucracy took umbrage at this interference in its affairs and never encouraged the development of Cyclades. As a result of this blunder, and the uneasy transition from Minitel to Internet, France still lags far behind other European countries in the use of Internet today. At this writing France Telecom still charges over $40 a month for a DSL connection, twice the amount users pay in Great Britain. As a result, fewer than 6 percent of French households have the kind of broadband connection that would permit the development of web-based businesses.

Other countries, notably Germany and the Scandinavian nations, also turned away from computer networking in the 1970s and 1980s, instead exploring interactive television as the way to provide information to the masses. The Prestel system in England was a typical implementation of that idea. Television had graphics, color, and speed, at a time when the early Internet looked decidedly dull, plodding, unreliable.

Dull or not, the growth of networking in the United States became irresistible. In November 1983 Paul Machapetris, helped by Jon Postel and Craig Partridge, came up with an improved system for e-mail addresses: the "Domain Name System," or DNS. As user groups argued endlessly about the abbreviations to be applied to the domains, like ".gov" for government and ".edu" for universities, there was a problem with the designation for those new interlopers into computer networking, namely, corporate sites.

It was Elizabeth "Jake" Feinler who cut to the chase with a momentous decision: company sites would be designated as ".com." At a summit meeting in January 1986, representatives of all the major networks adopted the DNS scheme, and the "dot-coms" were born.

By then the NSFNET far outweighed its Arpanet prototype, which Hafner and Lyon have aptly described as "a dinosaur unable to evolve as quickly as the rest of the Internet." Key researchers started leaving the government, taking positions in new structures. In 1986 Bob Kahn left ARPA to form a new nonprofit organization called Corporation for National Research Initiatives (CNRI), joined by Connie McLindon. CNRI started working toward a national information infrastructure. It was time to decommission the old Arpanet, transferring the sites into the regional networks. By the end of 1989 the Arpanet was dissolving into the general system of interconnected national and international nets. The last node was decommissioned in July 1990.

What survived, and kept growing exponentially, was the Internet. NSF had opened it up to commercial traffic, much to the chagrin of some computer science purists, and many companies were buying Cisco routers to implement their own internal networks. But communication across the net remained a technical challenge, filled with the drudgery of boring protocols. To the eyes of media professionals and communications experts abroad, the world of the Internet was still very dull indeed, hardly the stuff of a revolution for the masses.

All that was about to change in a dramatic way.

The World Wide Web

In 1999 I attended a presentation to entrepreneurs and business school students led by legendary venture capitalist John Doerr of Kleiner & Perkins fame. He gave a brilliant summary of the history of computers, ending with a description of the Internet. One hand went up in the audience. Was there any contribution from Europe? A young man asked. Well, no, said John Doerr, he couldn't think of any.

That was an amusing lapse for someone who had probably made more money than anyone else on Menlo Park's fabled Sand Hill Road investing in start-ups built on the concept of the World Wide Web.

The World Wide Web was invented in Geneva, Switzerland, at the CERN laboratories (CERN, which stands for "Centre d'Études et de Recherches Nucléaires," is a world-class facility for experimental and theoretical physics) by a young man, thirty-four-year-old Tim Berners-Lee. Frustrated with the difficulty of finding the research material he needed around the Internet, he invented a new language called "HTML," for "hypertext markup language," designed to help computers talk to each other and share data in more convenient fashion. The specifications for the use of HTML resided in what he called the "hypertext transfer protocol," or "HTTP." The whole thing could not work without a way of locating objects around the networks, so Berners-Lee also invented "universal resource locators," or "URLs."

The idea of hypertext (a term invented by Ted Nelson) was already present, in the form of "hyperlinks," in Doug Engelbart's first implementation of his system back in the 1960s. In fact, the original idea came from Vannevar Bush, who proposed it in 1945. What Berners-Lee did was to develop it fully by using the formidable leverage of the new interconnected networks. He started with "ENQUIRE" in the mid-1980s and did the first www implementation in 1989.

Throughout this book I have often used the terms "the web" and "the Internet" interchangeably, because they have become synonymous in everyday parlance. But this is only true as a gross approximation. The web is a superficial coating on top of the Internet, the outer multimedia structure that makes the network easily accessible to non-programmers through the browsers that now come with every PC. The World Wide Web encompasses those sites that have implemented Berners-Lee's protocols (HTML and HTTP) and make available a file structure that can be read by an ordinary browser (like Netscape Navigator or Microsoft's Internet Explorer) to users like you and me.

For most people, the web is the only aspect of the Internet that matters. But the Internet also contains millions of sites that have not chosen, or needed, to go through the transition to the web. They are the computing centers, the switching and communications engines, the data base depositories, the heavy-duty government and academic research facilities that serve traditional computer users or feed other machines that do not require a graphical interface.

If you need access to the source code of scientific models from NASA, or if you want to process data from the catalogues of the U.S. Geological Survey or the Lawrence Livermore Lab, a graphical interface would only get in the way of the process on your computer. These files are available through the old file transfer protocols established long before Tim Berners-Lee wrote his first line of code.

Setting aside the massive data structures of the traditional Internet, it is the development of the new graphical interface that has opened up networking to ordinary people. Once the CERN protocols were released, an embryonic web started forming among Internet sites, opening the way to even more sensational breakthroughs. In 1992 another young man, Marc Andreessen, who worked part-time as a programmer at the National Center for Supercomputing Applications in Urbana, Illinois, found a way to get graphics, photos, and audioclips through the web in a far more convenient way than had ever been possible with the most arcane computer software. Developed with his teammate Eric Bina, the program was called Mosaic. It was the first "web browser," a tool for navigating the thousands of computer sites linked together by the Internet.

My friend Dan Lynch, with whom I once worked at SRI and who went on to found the Interop show and serve as a member of the standard-setting Internet Society, told me of his colleagues' reaction when they heard Andreessen had released Mosaic to the public: "How could that happen? That young punk didn't even ask our permission!" was the shocked comment around the group. But a moment later, consternation was replaced by a sense of victory: "That's exactly what we wanted to happen when the net was designed! He didn't need to ask our permission."

The day had arrived when any bright programmer could come up with an improvement to the World Wide Web, and make it available to the whole world overnight. Some of these ideas would die, others would catch fire and revolutionize the planet. While there were about fifty commercial websites at the beginning of 1993, the number had grown to more than ten thousand by the end of that year.

In rapid succession, several such breakthroughs occurred. Jim Clark—with backing from John Doerr—founded Netscape with Andreessen and built the Navigator program; Brian Pinkerton released the first "web crawler," which he sold to America Online. The search engine industry had begun. Jerry Yang and David Filo launched Yahoo!

The world had changed. It seemed that the old dream of the Internet founders was close to completion. That dream had centered on information—how it was generated, saved, and shared, and most important, how the individual user would obtain what he needed from the vast pool of human knowledge online.

The Arrow of Information

Consider the following situation: You go out in the morning and pick up a newspaper. What you read has been preselected by a team of journalists and editors working all night long from thousands of dispatches, news items, photographs, stories, and reports. Only a small fraction of these items make it into the pages you will read. Along with the articles come advertisements geared to your community, ranging from a special sale at the corner drugstore to a major publicity campaign mounted by General Motors. Information items are being delivered to you like arrows shot at your door by a powerful army of skilled archers.

You may have the illusion that you have selected the way this information came to you. After all, it was your decision to buy that particular newspaper rather than its competitor. But your choices were very narrow indeed. If you had a specific interest in a subject the journalists did not cover—or in a subject for which the editors had no room in that day's paper—you will never get the relevant news through that particular channel. Nor is this observation uniquely relevant to the United States.

In countries like France, where advertisers have almost no constraints on their ability to invade public and private space (even plastering huge publicity placards on such cultural monuments as the

Louvre), the consumer is practically assaulted at every step. The French ad industry, backed up by politically savvy corporate giants, finds it easy to go around the law to promote restricted products like tobacco or drugs.[33]

If you rely on television for your information needs, the channel is even narrower. It may seem to you that TV, with all its fast-paced images and sounds, delivers a lot of stuff to your household in the form of news. But that is, in fact, a fallacy. Live television is a linear channel, which you are forced to watch by the minute without the ability to jump around, unless you resort to zapping, jumping from channel to channel, which only increases the confusion. With a newspaper, at least, you can scan an entire page at a time, your eyes focusing rapidly on what interests you. You don't waste time reading irrelevant commercials or silly classified ads. You can spot the financial news first, and keep the editorials for later. You can start with the comics, or the movie reviews. Not so with television.

Computer networks can change all that dramatically. They reverse the information arrow. *You* are the archer with the beautiful aim. You can point at the accumulated sum of knowledge out there and say, "Tell me what happened overnight in Bosnia," or "Did the Cubs lose again?" You can investigate obscure details of ancient history, ignoring everything else in the world until you get your answer. You can research the price of corn in Kansas or follow the most recent leads in connection with a series of crimes in Boston.

With the Internet, it is you the user who should be in control of the information channel, a feat made possible by resource sharing, the dream of the architects of the Arpanet. The user should be able to focus quickly on the most relevant program, the urgent page of text. It was the dream of Doug Engelbart, feeding filtered data to the fast-paced analytical skills of the human mind for concerned communities. And it was our dream at InfoMedia when we built conferences designed to bring experts to the cutting edge of their thinking.

The tools for the new arrow of information have been built. They are implicit in the structure of the web, in the browsers like Netscape and Internet Explorer, and in the search engines like Google, Yahoo!, Altavista, or Webcrawler that can pick out one paragraph of information out of billions of pages of text.

But the dream is not yet fulfilled. Essential elements are missing. The revolution has only just begun, and its victories are very fragile indeed.

One reason for the fragility of the new structures lies in the very scope of the cultural and spiritual changes they could precipitate.

In his book *Building the City of Man*[34] Warren Wagar remarked as early as 1971 that

> Because of Western technologies of transport and communication, the ecumene (the zone of habitation) increasingly becomes the planet itself. We begin to live on the scale of the earth. As Teilhard de Chardin pointed out, the earth's sphericity at last takes full effect; all movements of thought, all ways of life, all values thrust outward in all directions, meet, interweave, and fuse.

But Wagar also asked "whether it will be a mechanical synthesis of the essentially dead cultures of the old civilizations (my "solid state society") or a new organic system capable of further growth and expressive of living values (what I have called "the Grapevines"). But to prevent cultures from deeply interpenetrating throughout the world is impossible."

Teilhard himself had written that the mingling of cultures "impels us towards the ultimate formation, above each personal element, of a common human soul. . . . Everything that rises, converges."

Will the rise of the Internet lead the human spirit to such transcendence? Or will it eventually be seen as just another step in our long evolution toward a more connected species, with even greater inequality and injustice? Will the use of networks give us what futurists have called "a global brain for mankind," what Russell Targ has called "the first manifestation of planetary consciousness," or a more complex and chaotic world? The answer depends on the course the computer revolution takes in the next few years, and on our own actions as we become an active part of it.

6

An Unfinished Revolution

On January 14, 2000, Doug Engelbart gave a lecture at Stanford University on "The Unfinished Revolution" before an elite group of some forty-five information scientists recruited through the web. It felt like a return to the days of SRI, with a mixture of intellectual excitement and some confusion in the rush to new theories.

As an introduction to his lecture, Doug brought in his old friend Hew Crane to talk to us about the state of the world, the future of energy and oil reserves. Hew Crane, a mathematician and brilliant operations research pioneer, was full of the kinds of clever observations that make SRI such a wonderfully unique place. He set the stage with global statistics, pointing out the semiconductor industry was now turning out 10 million transistors per person per year, which amounted to more transistors than ants: there are about 6.2 billion people alive, and "only" ten million billion ants on the planet. Energy consumption figures were equally staggering: the world consumed a trillion gallons of oil per year, which was equivalent to a cubic mile.

This put the discussion of our industrial future in perspective. What Doug wanted to show by these statistics was the dramatic escalation of the problems faced by civilization, hence the acute need for new solutions based on information technology.

Wired for sound and firmly in control of his many PowerPoint slides, Doug stepped forward, looking as I had always known him, a man with overarching ideas and a sense of urgency in reducing them to implementation. This was all the more striking because of the setting, a magnificent multimedia amphitheater with video and webcast capabilities, multiple screens and online hookups at every desk. The presentation was sent out over the web to every point on the planet.

What impressed me most was that among the forty-five technically savvy people in the room, only one, a staff member, was using a laptop computer. Everybody else was taking notes on paper. I thought there must be a subtle message in that fact.

His sparse white hair carefully coiffed, Doug looked the part of the visionary as he recalled his online computer "library" of files and memos, the community journal and its complicated numbering system. While such a dynamic repository was dangerously off the beaten track at the time (an SRI vice president had told him, "IBM and DEC don't know anything about that shit," implying that the research was surely misguided and should be dropped!), the idea of an online record was hardly news on the Stanford campus of the third millennium. Doug's point that "something like this is surely needed in the world" drew blank stares from the participants old and young.

More relevant was Engelbart's claim that new technology could "boost mankind's collective capability for coping with complex, urgent problems." He added with a touch of humor that the purposeful pursuit of this goal was itself complex and urgent. Unfortunately, the solutions people proposed appeared to make the problem even worse. He attacked this by coming up with new terms such as DKR, for "Dynamic Knowledge Repository," and CODIAK, for something I missed.

To this were added ICs for "improvement communities" and ODs for "open document systems." What the world needed, he finally told us, was more NICs (Networked ICs) and particularly a multinational META-NICs. We were drowning in acronyms again. Jorge Luis Borges would have cried, or laughed, recalling his own story about the Library of Babel, which contains all possible books, including one book that explains the nature of the library, and one that explains why there is no such book.

After wallowing in what seemed a pursuit of purposeful complexity, I felt drained by Doug's vision. I was happy to drive home in the night, all velvety dampness along the Crystal Springs reservoirs, reality at last. Yet Doug Engelbart, one more time, had touched a very serious nerve with the unique perspective of his genius.

Harmony, Discord, and the Challenge to Authority

Whether you agree or disagree with Doug Engelbart, he forces you to consider his arguments in spite (or perhaps because of) their unfocused nature and their convoluted context. As I drove away along the coastal range toward San Francisco, I found myself thinking about Doug's observations, and holding a silent conversation with him in my head.

By itself, no technology can force people to accept agreement, to look for consensus, or to adopt a convivial behavior toward their colleagues. Networks have often highlighted elements of discord when they were expected to demonstrate greater group harmony. It is evident that one of the properties of the new media is to allow users to challenge authority with impunity, actually increasing conflict, revealing cracks in the structure. In the days of Arpanet, this last discovery had taken the Pentagon by surprise, but it was too late to put the genie back in the bottle.

Today governments that try to force regulation and make up new laws to control the Internet, its wild information patterns, its encrypted communication channels, and its overflowing electronic mail, are defeated by this universal aspect of the technology. We knew it as early as 1974, as reports of that era published by the Institute for the Future still show: while the network spreads, it allows human organizations to restructure themselves like a mad jungle, finding new ways around the traditional hierarchies with all the political and social consequences that implies. This was clear, too, in Engelbart's own writings of that era.

When any worker can send an electronic message to a vice president, or to the CEO of the company, bottlenecks are eliminated everywhere. The multiple management levels of traditional companies and the hierarchies they formed are replaced by structures of small autonomous groups who work in parallel and exchange data in a far richer manner.

The result tends to be upheaval, not smooth evolution toward consensus and rational problem solving. And the initial reaction, in many traditional societies, is resistance to the new medium. Nations such as Singapore and China have not embraced the Internet; they censor it. Even liberal France has sued Yahoo! because it objected to items sold at auction through the system, and it has long resisted any use of encryption by its citizens.

The French courts have gone as far as condemning, and throwing into jail, a poor individual programmer whose only crime was calling attention to the security flaws of the "Smart Card" widely used in France for everyday purchases. To demonstrate that the code could be cracked, he bought about ten dollars' worth of subway tickets without the system's registering a charge and told his story to the media, thinking he would be given some sort of medal. He discovered that he had committed an unforgivable offense against a financial scheme whose security was so bad that it could only be protected by preventing people from talking about its flaws!

It is a fact that networks can speed up productivity and increase the output and creativity of an enterprise, while reducing head count, providing greater span of control, and making conventional middle managers redundant. In a dynamic economy, this is a factor of progress, since new companies in creation tend to absorb the personnel who are freed up by the old companies that restructure themselves. But in a stagnating culture where ideology dictates the preservation of an obsolete employment system and where society remains under the control of an incestuous oligopoly, the web is seen to be as much a threat as a factor of progress. And in parts of the world where democracy has not yet penetrated, access to the Internet is viewed as a destabilizing influence.

The experiences of the last ten years contain useful lessons for today's Internet users. Of course, the technology has progressed with the number of servers and the density of websites, but one key factor has not evolved for a long time: *human nature*. While great speeches are made about technological revolutions and the wonders of new gadgets, the limitations and potentialities of the human brain have remained constant. The ability of human beings to receive, integrate, and process information obeys fixed laws, the same ones as in ancient Egypt or the Middle Ages. The technology of the Internet, even when equipped with some of the tools inspired by Doug Engelbart, has not changed this fact. Similarly, our relationships with others, whether through presence in a group, a voice on a telephone circuit, a picture on television, or a text on a computer, have stabilized around well-defined models.

Another striking fact highlighted on any chart of network development is the absence of major American corporations. Not only were IBM, Xerox, and AT&T unable to grasp the emergence of the Internet (when they were not actively hostile to it), but even the younger, more creative companies like Microsoft or Apple missed it and made no contribution to the building of the web. In the case of Apple, the appropriate time to design a network-savvy interface would have been the release of the Macintosh.

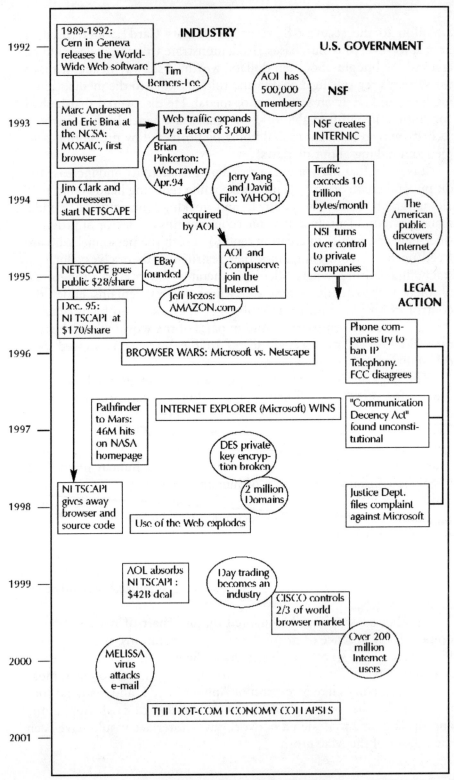

Chart 4: History of the Web

A colleague who worked for the company during that period explained to me that Steve Jobs and his designers had been inspired by the work done at Xerox-Parc on advanced interfaces, icons, and the mouse. All their energy went into the integration of these new technologies. But surely they could have stepped across the hall when they visited Parc and spoken to Bob Metcalfe, who had just invented Ethernet, a very fast protocol, as a way of linking computers together in local area networks. Or they could have learned from the experience with online communities at SRI and the Institute for the Future.

One of the readers of my 1982 book, *The Network Revolution,*[35] took it to Apple executives with the recommendation to capitalize on the emerging groupware market. He was sent away with the terse comment, "We make two things here at Apple: We make computers, and we make money."

The situation was a little similar at Microsoft when the web first appeared, with the difference that Bill Gates was smart enough to stay away from building hardware. With that smartness came a kind of arrogance that led the Seattle company to attempt to build and control its own version of the Internet. It only took three months for Gates to understand the mistake and realign his priorities.

Other industrial giants have not yet made the necessary adjustments to their business models. Entertainment companies, in particular, are still struggling with the impact of the web on the notion of copyright and free information exchange. The new media modify the context and deeply reframe social interaction. When they forget this lesson, our high-tech entrepreneurs and those who finance them make costly mistakes; witness the hundreds of millions of dollars wasted in the last decades on ill-conceived videoconferencing systems and on premature schemes for interactive television by sophisticated companies like AT&T or Intel.

Another concern is the strangling of innovation that results from expanding control of the web by media giants. In the words of Lawrence Lessig, Stanford professor and author of *The Future of Ideas*:

> The physical infrastructure is transforming as cable companies, and soon telecommunications companies, persuade governments to free them of traditional common-carrier responsibilities. As a result, companies can exercise more control over what runs along their wires and even decide what content flows at what speed, something called "policy-based routing." This change alters a crucial premise of the original internet. . . . By permitting such a fundamental shift, governments are allowing the enclosure of the information commons. That will destroy innovation.[36]

New Hopes, New Structures, New Spirit

It has often been observed that communication systems reflect the structure of a society to the same extent as architectural concepts do. My use of the word "structure" here is deliberate.

The first system that builders used for load bearing and stress distribution was composed of two vertical slabs capped by a third, a horizontal one, as we can see in Mayan constructions or at Stonehenge. The technique was revolutionized by the arch, and later by the keystone. By the time the ogival arch appeared in medieval cathedrals, always threatening to explode into pieces because of their taut construction, engineers had come up with arc-boutants, a technique that projected the entire structure upward. The modern stage of the builder's art is represented by the geodesic dome, where tension literally pulls the structure outward. In a geodesic dome, each polygonal segment is a source of both support and tension for its neighbors.

With each new step in technique, a new system of spiritual values was expressed, from the stark statement of the megaliths to the beauty of Roman structures to the elegance of gothic edifices and the all-embracing concept of the geodesic dome.

With each new step in technique, architects have been able to erect larger structures with greater span and integrity. Societal structures approaching the beautiful distribution of tension realized in the dome have yet to evolve, however. This is what network conferencing, in its finer forms, could realize. And here lies the difficulty in implementing it. Industrial society has remained at the hierarchical stage, with only occasional incursions into arches.

A geodesic organization would demand an information system that could instantly tie together all its segments when a collective decision was required; not in the sense of a broadcast or a mass rally, where masses of people are passively assembled at the receiving end of a giant, one-way channel; not in the sense of voice or videoconferencing, which demands the simultaneous presence of the parties at a few designated sites.

Instead, the geodesic structure requires an information medium in which participants can join the dialogue at any time, and in which past statements are not lost as the interaction progresses. It requires a medium where the arrow of information has been reversed. This is the medium Doug Engelbart was dreaming about. Computer conferencing, as we implemented it at InfoMedia, was a first step toward the creation of such a system: a revolutionary network where each node is equal in power to all others.

Today I still see networking as an ingredient for survival when the large and complex hierarchies of the past grow to such a size that they can no longer sustain the pressures they are creating. But I don't see it as a panacea. There is no "ultimate information machine," not even the human brain, whose limitations in memory and logic already have been made obvious by comparison with computing machines.

I like the idea of using groupware to facilitate new types of "grapevines" forcing old organizations to evolve. Informal networks and "invisible colleges" have always been the real harbor of trust and the spring of action for societies. What seems powerful to me is the combination of these grapevines—grown explosively to cover the entire world through casual connections made by invisible wires—with the resources already in place.

When the great failures of our human technological systems are analyzed, what emerges most often is not evidence of poor design or bad science. It is usually evidence of poor communication: the resources were there, but nobody connected them, nobody had a chance to ask the right question. The people with the inquiry never got the opportunity to talk to those with the facts.

With the advent of computer conferencing, there was a new tool to prevent another Three Mile Island, where the solution would have emerged if the scientists who knew the answers could have advised the managers trying to control the plant and the physicists trying to reassure the population concerned about a possible explosion.

The arrow of information needed to be reversed, with power turned over to network users everywhere. That was the dream of Paul Baran speaking of "building a better planet"; that was the dream of Doug Engelbart aiming at augmenting human thought; that was the purpose of InfoMedia with a ubiquitous medium for conferencing; that was the dream of all those who had participated, in one way or another, in building the Internet and developing applications for it.

But the revolution was not yet complete. The better planet had not happened. Now the world was manufacturing 10 million transistors a year for every man, woman, and child on Earth, and we were consuming one cubic mile of oil annually.

As they began to use the Internet massively for everyday commerce, information, and entertainment, millions of people became aware that the underlying technology was still very fragile. The networks on which we have started to rely for such an important part of our life are in their infancy.

Every morning the media report on problems people have experienced on the web: The auction site you planned to use to sell your grandmother's clock has been down for hours; your favorite stock trading site has experienced impossible delays; hackers in Poland have stolen another fifty thousand credit card numbers; and various user groups complain that their privacy has been compromised by bankrupt dot-coms that sell their customers' profiles to advertisers as their last remaining asset. Video downloading is so slow and fuzzy as to be discouraging; web telephony doesn't work very well; and online banking is seriously compromised by security concerns. Even such basic services as broadband DSL offered by major phone companies have been known to be down for ten hours at a time, or more. People who contracted with the telecommunications giant AT&T for cable modem service were cut off for over a week and permanently stripped of their online e-mail address by the failure of Excite@Home.

Nobody should be very surprised by these failures, because the Internet of today was never designed to provide all these things.

A Transition Technology

The Internet was a new network implementation but it borrowed most of its underlying concepts from the Arpanet, which was designed to demonstrate sharing of computing power and distributed databases across remote machines. That is why an understanding of the true history of networking is indispensable for anyone who wishes to make intelligent decisions about the use of the web. When Arpanet was built, the data were academic, economic, and military in nature. There were early experiments in voice and video as early as the 1970s but computer telephony, movies on demand, and financial transactions were not part of the mission; neither was security, nor data broadcasting, nor e-learning.

Contrary to the image of the Arpanet as a network of "Pentagon computers," most of the sites like our own at SRI were doing open research. As for user privacy, it was a joke: the point of the Arpanet was to share files and resources.

Now that we know what people want to use the web for, the entire network should be redesigned around a new set of goals. The problem lies in the fact that these goals are, in many cases, in contradiction with each other.

What we have today is a transition technology whose standards will seem absurd to the next generation of users just five or ten years from now. How will you explain to your teenage son or daughter that Daddy had to type: http://www.something.com

every time he wanted to visit a site on the primitive web? They will give you the same incredulous look I get from my own kids when I tell them I had to start my Renault 4CV with a hand crank on cold days when I was a student in France.

As French philosopher of science Michel Serres has reminded me, the current state of the Internet can be compared to the early days of metal ships, when the men-of-war had both steam engines and tall masts with sails.[37]

Technical changes are hard to implement once a product gets into real usage. Early in the development of the railroads in Europe, engineers realized the gauge they had selected as a standard was too narrow. Unfortunately, they had already laid down the enormous length of two hundred kilometers of track, so there was no opportunity for correcting the mistake! Trains have been using the wrong gauge ever since.

In the process of modernizing this bizarre contraption called the web, compromises will have to be imposed, in particular between privacy and security concerns. The problem is that the standards will not be placed at the service of folks like you and me, as they should be. In practice they are going to be set by an entirely new group of players with their own plans about information dissemination, the control of users, and the exploitation of network channels to maximize commercial profits.

And the dream is being compromised.

Part Three

The Betrayal of the Internet

> On a mountain, halfway between Reno and Rome
> We have a machine in a plexiglass dome
> Which listens and looks into everyone's home . . .
>
> —Dr. Seuss

Part Three

7

The End of Innocence

The Internet's age of innocence has been over for several years, although most web-based companies try to maintain the fiction of a network of universal access and free knowledge. Perhaps this idealistic picture was still believable at the height of the e-commerce bubble, when newspapers quoted Baran's expectation of "a better planet"[38] and T-shirts claimed that "INFORMATION WANTS TO BE FREE." But the collapse of the dot-com economy and the fight over the Napster music distribution system have reasserted the brutal reality of the marketplace.

Nowhere is the conflict between different visions of the future web more obvious than in the distribution of music. In recent years the major interests have brought into their empire the fledgling rebels of the music industry. Sony has gained control of MP3.com to initiate its Duet service, a joint project with Vivendi. At the same time a consortium linking AOL Time Warner, Real Networks, and Bertelsmann has neutralized Napster, the other would-be pioneer in the free-download era. As Wendy Grossman remarked,[39] "What the music industry wants is to make us pay as much and as often as possible for whatever it feels like letting people have."

Yet it is a joke to compare the cost of manufacturing and distributing disks or CDs and the cost of software dissemination, where it is the users themselves who pay for the equipment and the download time. As for the artists, they could be compensated more fairly under a system that cuts out the major studios, putting the material directly at the disposal of their fans.

What is true for music today will also apply to books, movies, newspapers, and television shows tomorrow, unless the new concept of peer-to-peer networking, which lets individual users link their computers together without depending on a central server, can win over the older structures.

Many interests are fighting to control the medium today. They have entered a new phase of systematic exploitation of the web. The Internet philosophy about user control is about to get compromised and betrayed by corporations that want to use the network purely as a distribution system. As Wendy Grossman points out, "There are many reasons to resist this Net-as-broadcasting scenario. The Internet's freedoms are what make it such a useful way of exchanging information. There is no way to control what people do with the files they download without invading their privacy." The recording industry is already trying to monitor the use of Gnutella, a Napster alternative, by detecting who has what files on their hard disks, and threatening them with lawsuits.

The newest tactic of music studios is to flood the file-swapping networks with entries claiming to be music, containing only a short piece of the promised song, followed by noise. Corrupt the data, and the database is useless!

It doesn't have to be this way. All of us, as individual users, retain ultimate power over the information we select and use every day. But to exercise this power will take much greater sophistication and awareness than it did in the previous era when the web was open.

A Dual Threat

Two major threats are developing against Internet users today. The first one comes from the advertising and marketing industry trying to influence our behavior as consumers: a reversal of the information arrow. The second kind, which we will describe in some detail in chapter 9, comes from governments eager to control what people say, read, and think.

These two threats have a lot in common. To begin with, both reflect legitimate interests. The marketing of new products demands an understanding of the tastes, desires, and needs of consumers. The Internet is an ideal medium to find out what they are, so any corporation that deals with the public has an urgent desire to obtain this information.

Governments, too, need to know something about what people do on the Internet. The web has become a magnet for all sorts of criminals ranging from pedophiles to swindlers, money launderers, and terrorists. Clearly these activities should be detected, monitored, and curtailed.

Many people are reacting to this state of affairs with a form of paranoia, either resorting to elaborate encryption systems for the simplest exchange of e-mail, or staying away from the use of the Internet altogether. Such fears are inappropriate, and only lead to paralysis in a society where web-based services are becoming increasingly important and convenient.

Unless your first name is Osama, the CIA is probably not interested in knowing how many pieces of electronic mail you send your friends every day; but AOL certainly is. Similarly, unless you are extremely interested in explosive substances, the FBI doesn't care what kind of book you order from the web; but Amazon has already cross-indexed you with dozens of titles they believe you should read next. The same goes for music and films, canoes and golf balls, movies and airline tickets. Vivendi, Thorn EMI, United, and Disney all want to know how you are likely to react to their next ad or the price of their products.

The problem comes not with such legitimate applications, however, but with the way in which they are abused. Advertisers who could only get aggregated statistics on responses to their campaigns can now track our personal habits undetected, gathering data about our purchases and even our future intentions as we browse various websites looking for books, gift ideas, vacation plans. Government agents can snoop on us in many ways that were unavailable, impractical, or plainly illegal before the Internet. They can insert filtering software on the computers of every Internet service provider (ISP) company. Using their own computers, they can analyze individual user patterns with the help of massive databases.

Occasionally the stupidity of online data-mining reaches a point of comic intensity. Thus a woman named Cathy Watson reported that she had been searching the web for articles on fetal distress when a series of ads popped up on her screen: "Lowest prices on *fetal distress* at DealTime. Compare prices from 1000s of Stores and Save Instantly! Look for *fetal distress* at eBay—the World's Online Marketplace!"[40]

The simplest example of a technical tool that can be used to find out who accesses what information, when and why, is called a "cookie." It is a short string of data written onto the hard disk of your personal computer upon request from a remote service. The cookie file is not intended as an intrusive device, but as a way of helping users surf the web. It is used to facilitate your interaction with the same service at a later date. But it can also serve to determine what kind of information should be presented to you in the form of goods, services, or simple advertisements.

So what?

In principle, there is nothing wrong with the idea of optimizing individual users' interactions with common services. And it is a good idea to redesign or expand the web to make it more reliable, more efficient, capable of supporting new media, and offering innovative companies a chance to thrive economically. The world certainly doesn't need a repetition of the dot-com debacle that deprived us of some valuable sources of data and many good services. You may even welcome some forms of advertising tailored to your interests.

The problem is that, in the process, the basic notion of the Internet as a universe of liberated intellectual exchange and increased communication freedom is radically compromised. The web is being turned into a mass medium like all the others, with the information arrow pointing the wrong way, and control firmly in the hands of the marketing companies.

Some services try to anticipate their customers' desire for disclosure about the data they gather. Thus, as a user of Netscape through Pacific Bell, I received an e-mail notice that the company did not share user information with other organizations. If so, I replied, why was it that my browser was always detecting a series of requests to write cookies on my hard disk, coming from a remote server called "ads.web.aol.com" every time I loaded Netscape?

An anonymous message came from Pacbell a few days later, instructing someone in the SBC Internet Services to respond to my inquiry. Indeed, a fellow called Andy (or perhaps a software robot masquerading as a fellow called Andy) followed up. I could contact the SBC Policy Department to report hacking incidents, spammers, and virus propagators, said "Andy." He even gave me a number to call. But he did not seem to realize that I wanted information about what looked like abusive practices from his own company! At the very least, I should be able to turn off the requests from ads.web.aol.com, whoever they are. I imagine they are related to Netscape, and simply log what I've seen in the course of a session to get paid by their advertisers. But they slow down my access to the Internet, and have no need to know at what time of day I use the web, or for what, without my permission. I turned off the cookies.

"No cookie, no more personalized browser pages," Pacific Bell told me. I have not used Netscape since that day, except as a path to my e-mail. Fortunately, I have access to another service that doesn't demand cookies. How long can I stay one step ahead of the Solid State Society?

Most users would never even know that such services gather information underneath the surface of their web transactions. I became aware of this particular one only because I am curious, and make it a practice to turn on a browser feature that alerts me to such cookies. (See the section "Countermeasures" in chapter 10.)

A Pacific Bell technical service person I once called for advice told me that many people were actually shocked when they reported some obscure problem and she asked, "When's the last time you've cleaned up your cookies?" Admittedly a peculiar question to ask a stranger in the course of polite conversation.

The Golden Age Is Over

The year 2000 can be taken as a useful marker in the deployment of the web. At that point in history there were 37 million hosts on the Internet, 300 million websites, and 72,000 newsgroups. It was estimated that in July 2002 there were 577 million online users around the planet.

One can quibble with these numbers. Perhaps the estimate of the total number of users should be divided by 5 or 10. For instance, I have several redundant identities on the Internet through various websites (the simplest way to reach me is through the e-mail address documatica@aol.com), and my daily usage of the network is fairly average. Most people have separate identities on their employer's internal network as well. They log in with a different name from their home machine. Yet the fact remains that network technology is growing much faster, and penetrates human society in more decisive ways, than any previous invention, be it the telephone, television, electricity, radio, or the automobile.

The Internet also looms huge as a marketplace, with an estimated revenue figure likely to be in the trillions of dollars within a few years, even after the spectacular crash of the first generation of primitive e-commerce sites. Usage of the network did not decrease during the period when thousands of dot-com companies failed. Only 8 percent of American users saw one of their favorite websites go out of business. Of that group, only one-fifth were spending less time online, according to a survey by the Pew Internet and American Life Project, published in July 2001. Half of the future electronic commerce traffic is expected to be international in nature, and content will be king. The pressure to monitor and control user behavior in this environment will be enormous.

Given these astounding numbers, how can our freedom to communicate be constrained, rather than expanded, by the new technology? The answer has to do with that subtle law of cybernetics that claims that information is control.

There was a blissful period, between the release of Mosaic by Andreessen in 1993 and the dot-com crash of 2001, when Internet users experienced personal control of the information arrow. They could determine what they wanted to see, when they wanted to see it. Search engines put immense power at their fingertips. The answers were not always accurate, but human language itself is ambiguous. The search engines would plow through enormous volumes of data, almost in real time. You could "spider" through much of the fine content of the web. From then on, you could explore to your heart's content.

Many important things happened in 1995. RealAudio launched the first product for audio streaming; Netscape went public at an incredible valuation. Amazon revolutionized book distribution. The year 1995 is also when the National Science Foundation decided it would no longer provide direct access to its backbone. Private companies took over access management and started charging for domain names.

The only people who complained were the old-line computer scientists who had regarded the Internet as their property. They felt outnumbered by the "newbies" and looked down their academic noses at anyone who didn't know a universal record locator from a secure socket. For the rest of us, the rise of the Internet during that period was exhilarating. One made discoveries every day. There was so much fresh experimentation that you could literally spin a virtual wheel and land on remote sites at random to discover what they did. New ideas, programs, and techniques spread around the planet in mere days.

All that changed when commercial reality reasserted itself. About the year 2001, the arrow of information tended to reverse. Without most users paying attention, control began to pass into the hands of a new generation of sophisticated companies that are wielding power in the digital society.

As of early 2003, 50 percent of all books published in the United States come from only seven corporations. And 50 percent of all web surfers visit sites that are owned by just five major groups of companies. So much for the spread of technology to encourage intellectual diversity. Under the appearance of increased freedom, what is actually happening to us, as users of information, is not loss of virtual opportunity but effective loss of control.

Advertising Abuses

In their early incarnations, Internet ads were crude banners characterized by their vivid colors and blinking intensity. The dot-coms of the first generation made money by selling space on their pages to advertising companies. The practice became so prevalent as to reach absurdity. Much of the industry tried to support itself by cross-advertising on each other's websites, setting up a house of cards that was ripe for collapse at the first sign of a slowdown in usage. A pet food company would advertise on the site of a car dealer, and the car dealer would announce his latest rebate plans on the pet store's own page, in a completely incestuous system.

We still see unsolicited banner ads on our computers, but nobody pays much attention to them anymore. Many firewalls and virus checkers will screen for such offensive material and delete it. But the commercial targeting of your personal profile is now happening in far more subtle ways.

In the example of unsolicited cookies triggered by the use of Netscape, I know that I can proceed with my session even if I deny authorization for the advertiser to pollute my hard disk (although some systems "hang" if cookies are blocked), but the very fact that my profile can be accessed in this way, and sold to unseen parties, makes me nervous. It can determine what news is fed to me, what new services are offered. Even though limitations have now been placed on the abuse of cookies, their very presence skews your usage of the web in ways so subtle, you cannot always detect them.

Senator John McCain, the Arizona lawmaker who heads up the Senate Commerce Committee, opened hearings in June 2000 with the observation that "through the use of cookies and other technologies, network advertisers have the ability to collect and store a great deal of information about individual consumers. They can track the websites we visit, the pages we view, the duration of our visits, terms entered into search engines, purchases, responses to advertisements . . ."

You may have the impression you are free to roam the Internet at will, but there is an invisible veil between the whole process and you. Watching the interaction. Biasing the results.

Often the bias is not so subtle. In a bold move to place increased revenues above user privacy, bookstore giant Amazon.com announced it would make previously protected buyer profiles available to whomever it wanted. If you have ever bought a book on breast-feeding, don't be surprised if you still get offers for discounted baby products years from now, when your kids are graduating from MIT. If, on a whim, you have ever purchased a risqué novel involving some adulterous affair in far-away China, don't complain when your nine-year-old daughter is blasted with a screenful of advertisements for bodice rippers when she borrows your PC to do her science homework. And her innocent request for chemical data on nitroglycerin could bring the FBI to your home the next time someone tries to blow down the back door of your local bank.

Writing in May 2001, Brad Grimes reported[41] that Toysmart attempted to sell its database of customers when it filed for Chapter 11 protection in June 2000. The site's privacy policy specifically forbade this practice, yet conflicting laws led to this violation in spite of suits by the FTC and the attorneys general from forty states. "Eventually, after receiving $50,000 from a division of Disney, a financial backer of the now defunct retailer, Toysmart.com agreed to destroy its list," writes Brad Grimes.

Another even more offensive form of abuse was instituted by the auction company eBay. In January 2001 it revealed to its users that the default for user surveys asking questions such as "do you agree to receive marketing calls?" had been mistakenly set to "no." The company restored the preferences to "yes" and gave users two weeks to change the answer if they didn't agree, but it was now the user's responsibility to block the intrusion.

Many services, such as banks and insurance companies, similarly indicate in the fine print of their contracts that you are letting them share your personal information with other firms and advertisers, unless you specifically forbid this practice. Thus I have before me a notice sent in August 2001 by Carolina First Bank, an institution that supports business online, stating, "Unless directed otherwise, we may share all five types of customer information within our family of companies. We are permitted by law to share information about our experiences with and transactions with you or your account."

Fair enough. But the statement goes on, "We may also share *other customer information* [italics added] that is not considered to be account or transaction experience, including your information obtained from your application (such as marital status or income), cus-tomer reporting agencies (such as your credit history) or other outside sources (such as employment history)."

In other words, this bank, where I have never set foot, can send confidential information about me to anyone it considers to be in its "family." I can direct them not to do it by filling out and mailing a spe-cial form, but most people do not even read such notices, and few of those who read them will go to the trouble of filling out the form and mailing it back. Which is exactly what these institutions are counting on so they can happily go on exploiting your personal data for their marketing.

Have no illusions that information remains contained within a "family of companies." This particular bank belongs to a family of nineteen firms. Many groups are much bigger.

In November 2000 it came to light that French cable TV and pro-duction company Canal Plus, which maintained a file of 4.5 million subscribers, would make it available to its affiliated companies within the Vivendi group. Writing in the French daily *Le Figaro*,[42] former budget minister Michel Charasse complained that none of the organi-zations claiming to protect privacy in France seemed to be clamoring against such abuse. Vivendi controls a major utility company (Générale des Eaux), a telephone company (Cégétel), and France's leading adver-tising agency, Havas. It is a major player in cellular phone and Internet access, and owns Universal Studios in the U.S.

Michel Charasse stated that he would personally go to court "if they give my name and address to people who don't need to have them and are not authorized to use them, especially to increase their market shares and make me a party in spite of myself to the prosperity of the Temple's merchants."

The same practices go on in the United States, in spite of citizen groups' efforts to control them. The magazine *Red Herring* reported[43] that a little-known company called Acxiom, in Conway, Arkansas, had amassed dossiers on 160 million Americans, representing 90 percent of U.S. households: "Acxiom has 20 million unlisted telephone num-bers—gleaned mostly from those warranty cards you filled out when you bought that new coffee maker—that it sells to law enforcement agencies," wrote the magazine. It also sells it to lawyers, private inves-tigators, debt collectors, and anyone else who pays for the information. The largest data-mining company in the country, Acxiom has annual sales over $1 billion and 5,600 employees worldwide.

An example of the use of its data integration software, called AbiliTec, is the ability of Mercedes-Benz to obtain detailed data about shoppers whether they are online, on the phone, or at a dealership. The salesperson at Mercedes can assess the profile of the caller in real time.

It is noteworthy that data-mining research, generally defined as the development of tools to employ keywords to build a database or bases from a number of large databases designed for uses different from those required by the analyst, was pioneered by the intelligence community in the U.S., notably the Office of Advanced Analytical Tools (OAAT) of the Central Intelligence Agency.[44]

Where Vannevar Bush Went Wrong

Today I looked at my late-model laptop and wondered what Vannevar Bush would have thought of it.

This amazing piece of electronics has all the attributes of his "Memex." It is the ideal personal assistant, with a precise recollection of my contacts, their addresses and phone numbers. It can keep track of my schedule, remind me of appointments, draw maps to get me there and keep my messages organized, with regular updates through wireless links. It has a capability for computation, budgeting, and accounting that is extraordinary, and it can also help me compose text, such as the manuscript of this book, without getting confused about successive versions, wherever I store them away.

Even better, this machine is aware of my own occasional distraction and helps set me back on the right track if I call up the wrong file.

It also differs from Memex in several important respects that detract from its usefulness to me. These differences were not anticipated by Bush. They reduce the trust I am tempted to place in my assistant.

To begin with the most obvious observation, this laptop does not really watch out for its master's interests or even respond to my commands in the way I intend them. The various pieces of software that make it tick do not belong to me: I am only borrowing them. These systems belong to corporations like Microsoft and AOL-Time Warner that use it to influence my actions. They hide some information from me, shield me from opportunities they regard as competition, and keep interfering with my wishes.

"You are connected to the Internet. Would you like to start AOL now?" asks my assistant with spontaneous obsequiousness.

The Memex would never intrude in this way. There is an icon for AOL on my screen. If I wanted to start AOL, I would click on that icon. I may be distractible, but I am not stupid.

This assistant of mine also suffers from a perverse type of multiple personality disorder. Obscure pieces of powerful software seem to be fighting for control inside my machine, attaching themselves to files and connections, grabbing space on the disk without asking permission, dragging in their associates, downloading irrelevant stuff when I'm not looking, and elbowing each other to be the first to respond when I try to connect to the outside world. Among the worst offenders are RealNetworks, which conflicts with Windows media products, and Microsoft's camera services that try to steer you to their own partners for picture printing.

These pieces of software think they own my laptop as a display window for their own wares. Throughout every session, they pelter me with ads and spurious e-mail. Microsoft's Internet Explorer pleads with me to elevate it to become my default browser, to the exclusion of lesser programs. The "INSTANT MESSAGING" icon of AOL has never gone away in spite of all my attempts to expunge it. AOL instant messaging is incompatible with software from competing companies, yet it keeps intruding into every session of Netscape. Similarly, Yahoo! interrupts the flow of my thoughts with screen after screen of "helpful" offers for starting my webcam, or putting me in touch with folks it has, perhaps mistakenly, identified as my "buddies."

I have to keep saying *no!* to every one of these helpful suggestions before I can proceed with my work.

These problems are magnified with the most recent (at this writing) release of Windows, the XP version. It keeps trying to get the user to sign up for Microsoft's Passport system. While Passport eventually will provide unified access to multiple services from the company, it is not necessary to subscribe to it in order to use Windows XP. However, it won't leave you alone until you have ignored the request no less than five times! Even then it remains as an icon at the bottom of your screen, ready to spew out more Passport ads at the first opportunity.

Dr. Vannevar Bush would throw away this perverted implementation of his Memex, and go back to the drawing board.

The Search Engine Bias

In July 2001 a consumer advocacy group called Commercial Alert filed a complaint with the Federal Trade Commission accusing eight popular websites of deceptive advertising. The sites included Microsoft, Altavista, Netscape, Lycos, Hotbot, and Looksmart.

There is a thin line between search engine results and plain advertisements, and that line is often trampled on the Internet today. The search engines do not always disclose the fact that the results they present to you are often paid for by companies, although they are now under pressure to reveal this fact by identifying paid results as "sponsored links."

These companies actually buy keywords that the indexers sell to the highest bidder. And sponsored links always appear on the first page of results.

Suppose you have just been diagnosed with a certain disease. You are searching for medication data on the web. Assume that a researcher at an East Coast medical college, Dr. Plunkett, has written a remarkable thesis about the ailment, and has posted it on her website. But Big-Pharma, Inc. has its own ideas about treatment, based on its multibillion-dollar drug, Cryptonase. What makes you think that "Plunkett.com" will come up first on your search engine? Cryptonase has paid millions for the privilege of being at the top of the list.

The justification for selling indexing to the highest bidder is a lovely example of greed in action. It claims that sites that have the economic wherewithal to pay for keywords should come up on top of the list because their great wealth reflects their importance to society. Is this necessarily so? Is that the way we will find out about fragile new ideas? Doesn't that violate the very notion of an information universe where we, the users, are in control of what we see? Whatever happened to the idea that we could move across the various delicate layers of the billions of pages put at our disposal by website authors?

No one knows what the FCC will do with the complaint. In the past the agency ordered television stations to identify infomercials as advertisements, but it hasn't yet come to grips with the more subtle biases on the Internet. The result, however, is already evident: the Internet's usefulness as a research tool has been severely diminished in the last few years. And the people who only look at the first few entries listed by their search engine are looking at advertisements, not genuine information.

In a recent twist on this notion, a San Francisco company called Ezula introduced an online "contextual advertising" technology that turns ordinary English words into hyperlinks without the user's authorization. According to journalist Benny Evangelista,[45] this "feature" first appeared on the screens of users of an online file-sharing program called KaZaa in July 2001. When they installed KaZaa on their computers, they also downloaded Ezula's program called TOP-text, which turned some words like "jazz" or "hip hop" into hyperlinks.

When you touch these words they send you to the website of an advertiser, such as a music publisher.

Companies buy the words they want to highlight, and they typically pay Ezula fifteen cents per click when users follow the hyperlink. There is no limit to what you could do with "contextual advertising." You could buy the words "travel" and "holiday," making sure that any potential vacationer would be offered your packaged tour of the Great Pyramid. You could also purchase words such as "God," "race," or "politics."

Ezula's cofounder told a reporter, "We finally can deliver the promise of advertising on the Internet." Other people see it as a form of insidious and annoying violation of users' rights to read unaltered text, and to define what links they want to follow rather than predetermined paths to a commercial site.

The RealAudio links are especially annoying. Every time you need to use their product they take you to their own website for advertisements and updates, whether you like it or not. My son, an analyst who frequently needs to listen in on company conference calls, finds the delay obnoxious: "It's as if your car was programmed to drive you over to the nearest Radio Shack every time you turn on the news," he observed. Yet there is nothing you can do to stop such abuse.

So far contextual advertising has been distributed mostly on music sites. But the practice could spread everywhere, just as banner ads once did.

Yahoo! announced in 2001 that it would propose targeted ads to users, based on the country and the city where they were located, using a technology of geographic tracking Yahoo! obtained from Akamai.

This move was ironic, because just a few months earlier the same Yahoo! had told a French court that it was not able to filter out users from France who sought to buy Nazi memorabilia in defiance of French antiracist laws. (In November 2001 a federal judge in San Jose, Jeremy Fogel, ruled that Yahoo! didn't have to obey the French court's order. The company had successfully argued that censorship of third-party content on the Internet violated the First Amendment.)

Another site is trying to go even further than Yahoo! with its targeted ads, according to the *New York Times*' Paul Krugman. Shopping on Amazon, he noticed that the company was charging different prices for movies, based on the customers' profiles. This practice, reportedly known as "dynamic pricing," appears to be an electronic version of price discrimination, charging a premium to a potential customer based on his or her record of purchases, and on the associated address.[46]

I might as well confess that in spite of these abuses, I still love the Internet.

I love it because it lets me listen to music and news from any place in the world. Even in San Francisco, a city that thinks of itself as highly cultured and sophisticated, the fact is that the major newspaper squeezes world news in the back of its obituary section, while the single surviving classical FM station plays Mozart all day long ("So relaxing on the way to work!" says the mellifluous announcer). If you are anxious to know what goes on in Brazil or England, and would rather not listen to *Eine Kleine Nachtmusik* for the twelfth time, the web is the place to go.

The Internet also lets you do crazy things, like visiting the site that teaches you the Klingon language or listening to the international news in Latin from a station in Finland.

As you do that, however, remember that hundreds of monitoring servers are watching. Databases are being updated as you switch from Beethoven to Shostakovich, as you order a copy of Hamlet in the original Klingon, or consult the timetables for the ferryboat in Dover. Anything you do online may affect your ability to get a favorable rate on a mortgage, or a deal on that new Mercedes.

On the web of the future, we are more likely to feel like the fly than the spider.

8

Building the Future Mesh

In the days that followed the World Trade Center attack in 2001, the FBI subpoenaed computer records from major Internet services and from several public libraries on the East Coast, looking for evidence that the terrorists had planned their strike using e-mail. Indeed, a librarian in Delray Beach, Florida, confirmed that Marwan Al-Shehhi, believed to have been one of the hijackers of United flight 175, had signed up to use a computer there. It appeared that Bin Laden and his associates had also planted maps and other coded instructions on X-rated websites. To unravel the techniques used by such a group to encrypt its data will take a team of exceptional skill and a great deal of luck.

As early as the 1940s Alan Turing had foreseen the day when computers would become so complex that no human being would be able to determine how they performed certain functions. The interconnected parts of the Internet today represent such a system. In the winter of 2001 the network of the National Security Agency, the most sophisticated group of computer scientists in the world, was down for three days. Not as a result of an attack or a virus, but simply as a result of data congestion that brought down one part of the network after the other.

"Fortunately it happened at a time when Washington was rather quiet, and not much was happening in the world. Many federal employees were kept away by bad weather; so few people realized the flow of NSA intelligence had been interrupted. And there was no single person who understood the whole network well enough to bring it back up," said the security executive who gave me this information.

With the spreading of the web, a new planetary being is arising, a wondrous and dangerous entity made up of collective thoughts, far-reaching sensors, and all the accoutrements of artificial life. While some writers have hailed the web as "a new brain for mankind," it could be more appropriately compared to a new nervous system, with all the complexities, false alarms, and unlimited opportunities for both accurate warnings and biased information flowing into our collective consciousness. In a magical sense, it represents an instrument of evocation where energy and information are intertwined in ways no physical theory could have anticipated.

It wasn't that long ago that the engineers at Bell Labs solemnly announced that the upper limit for the information that could ever be transmitted down a copper wire, according to Shannon's law, was 1,400 bits per second. Most households in America today break that supposed law of physics every day, by two or three orders of magnitude.

From the Web to the Mesh

The web marked the apotheosis of the twentieth century's "information age." As we make our first uneasy steps into a new millennium, the network has become so strange and so difficult to describe that one is tempted to talk about it in near-mystical terms. As an esoteric instrument to peer into the past and future, it is more powerful than the scrying glass of Doctor John Dee, more subtle than the water-divining vessels of Nostradamus, more mysterious than the athanor of Nicolas Flamel. Yet it is mired in more noise, and is more encrusted with false information and misleading recipes than all the grimoires of nefarious sorcerers. It has more blind alleys than alchemy.

In talks and interviews given in 2000, Tim Berners-Lee, the conceptual creator of the web, expressed worries that it would grow out of control as commercial developers piled layer after layer of new software on top of the simple structure he had designed.[47] He had first proposed the web in 1989 while working at CERN on problems of network control. The project was never formally approved, but he worked on it quietly with his manager's permission. When an information retrieval system called "Gopher" appeared, it seemed to be a competitor to the web, but the University of Minnesota stupidly tried to charge for the software, and Internet users dumped it in 1993. Then a team from the University of Illinois came up with Mosaic, the first web browser, assuring the survival of Berners-Lee's invention.

Berners-Lee said that the beauty of the web was its vast potential for spreading knowledge. Another quiet man who coauthored the "World Wide Web" proposal at CERN, Robert Cailliau, observed that "it's simply not possible" to guide the way the web develops: "One should not lament over the property that the web has—being adaptable to all tastes."[48]

An Evolving Culture of Information

The field of computer networking is now old enough to produce its own culture, complete with a new mythology and folklore. In the process, the record is getting blurred, and some really absurd claims are being made.

One of these claims is the often-heard statement that you can find everything on the World Wide Web. This is plainly false. The Internet was not even designed to find everything, nor is it necessarily desirable to have a network where you can find everything.

I cringe every time I hear Washington politicians boasting that soon all the children in America will "have the entire Library of Congress at their fingertips." I wish it were that simple. It takes a two-hundred-page book just to describe the index scheme for the Library of Congress. Students could take the better part of a year trying to decipher it.

It is certainly desirable for the Internet to be available from every school, but the benefits to education will not be any more automatic than the computer education systems of forty years ago I mentioned in the first part of this book. Like the Plato system in the old days, e-learning on the web will require intellectual and affective guidance from adept teachers and parents in order to be effective.

Contrary to repeated claims by every administration, accompanied by carefully framed snapshots of Bill Clinton and Al Gore in their shirt-sleeves hooking up Ethernet cables in ghetto institutions, Internet access from American classrooms remains small. In 2000 only 13 percent of students accessed the Internet from school as well as from home. Those who only used it from school (presumably because there was no computer hookup at home) represented 1 percent of all American kids. The majority of their peers did not use the network at all. So much for having the Library of Congress at their fingertips.

You do find an extraordinary number of things on the web, but they are drowning in junk, again reminiscent of Jorge Luis Borges's story "The Library of Babel." Such exaggerated claims are part of the betrayal of the Internet, because they promise something that cannot possibly be delivered. Worse, they set up our children for another source of bias in an education system already fraught with blind alleys and subject to many misleading notions.

Next: Building the Grid

So, where is the Internet going? To help myself answer this question, and at the same time clarify a few points about the early history of the technology, I called Paul Baran and arranged to have lunch with him and my friend Steve Millard at a quiet Japanese restaurant in Menlo Park. The timing was perfect, because an article had just appeared in the *New York Times* by Katie Hafner examining the claims of computer scientists who were arguing about who should get credit for inventing packet switching.[49]

After building a string of successful publicly traded companies (from Equatorial and Telebit to Stratacom and Com21), Paul Baran has earned a comfortable lifestyle among established Silicon Valley digiterati and, as he puts it with a smile, "I don't need another plaque on the wall." So he is looking at the disputes around packet switching with an amused indulgence.[50]

"My work on all this goes back to my first distributed network proposal at Rand in 1960," Paul said. "I used the term 'message block' in the early 1960s. In 1965 Donald Davies in England, unaware of my earlier work, independently came up with the same basic concept. He chose the same data rate as I did, at 1.54 Mbps, and the same length packet of 1,024 bits. Davies called his system 'packet switching,' a better choice of words."

"Davies died recently, didn't he?" asked Steve Millard.

"He died just before he was going to receive the IEEE [Institute of Electrical and Electronics Engineers] Internet award, in June 2001. His son received it in his name."

"Is it true that the motivation for the Arpanet was to decrease the vulnerability of military communications, or was it a research need, as the Hafner and Lyon book about the role of Bob Taylor has suggested?"

"Both are true," Baran said. "ARPA was sponsoring computer research around the country, and there was a serious lack of coordination. The first idea, if you can believe that, was to move all the computers to a single room in Omaha! That's when Taylor and Licklider thought there must be better ways to do it, and came up with the idea of leaving the machines where they were and building a network. As early as 1962 Rand had asked me to think about novel ways to preserve communications in a nuclear exchange, when collateral damage could take out all our systems. The real danger was accidental stupidity, a nuclear blast that would affect the ionosphere, take out all radio links and damage the phone equipment. The problem was to find ways to survive the first strike and still control the second one. So we came up with the concept of 'minimal essential communications.'"

"The whole phone system was analog in those days," I said.

"Yes, and the phone companies had expensive links from one end of the country to the other. They didn't think there was any other way to do it. That's where we ran into problems with them. I tried to show that, if links are cheap and numerous, an interesting thing happens: cost goes way down, of the order of two hundred to one, and you are no longer vulnerable to a single failure. But for that to happen, the system has to be digital. And you have to build it with two or three times the links you would need under their old scheme. That's the system that was finally implemented in the Arpanet."

"How do you see the network evolving now?" I asked Paul Baran.

"Cost will continue to go down rapidly, as we move to DSL and other broadband systems. And the future architecture is peer-to-peer computing, where end users can tap the power of the entire network without relying on central servers. The cost of the last mile is the big remaining problem."

Most experts agree with Paul Baran that the major strength of the web resides in its ability to grow organically. Grow, it certainly will. Like the deity worshipped by the Yesidis of the Middle East, it is a god that duplicates itself, existing at many levels simultaneously. And the benefits for us, as users and consumers, are very obvious.

In the words of Ethernet pioneer Bob Metcalfe, now a venture capitalist at Polaris Venture Partners in Waltham, Massachusetts, the network will bring increased prosperity: "As access and transaction costs come down, it will be easier to find what you want, and you'll be able to get it cheaper."[51]

Dr. Larry Roberts, the brilliant manager of the Arpanet project who deserves much of the credit for the early development of networking, projects the future in terms of improved quality of service.

His website points to 2001 as the year when guaranteed rate of service for voice and video was achieved, and it expects that quality and security will have arrived by 2005 to the point where "there is a major trend of both home and office users to use the Internet for their primary voice service, their high-quality TV and radio service, as well as their data service."

Roberts expects that "this trend will slowly eliminate all the other communication nets like the current telephone network over the next decade."[52]

The current limitations have to do with exactly how communication does happen. In that respect, two things are noticeable in the usage of the network today.

The first one is a crisis in bandwidth allocation. Simply adding phone lines, fiber optics, and wireless transmission will only make the problem more acute, like the "turnpike" phenomenon in urban areas. The bigger the turnpike, the worse the traffic jams. In a world where users have not even begun to routinely send video to each other, the bottlenecks are already so severe that you may have to wait for a page just as long on a DSL setup as you did with your old, slow dial-up modem. Access points to your metropolitan area are clogged up like old pipes. As Larry Roberts points out, the solution will come from better protocols for bandwidth allocation and corresponding payment rates, an economic reality the major telephone companies have severely underestimated.

By the way, DSL (which stands for "digital subscriber line") wasn't even designed for faster Internet access. The phone companies conceived it to support interactive television, their old obsession. Then the Internet market took off, and other companies failed to provide bandwidth. Cable companies were slow in deploying modems, and two-way satellite links had their own problems. It was a fluke in the adoption of technology that made DSL the standard it is today.

So much for the technical stuff. The second trend in evolution will be qualitative rather than quantitative. The architecture of the web, based on access points maintained by the phone and cable companies, connecting you to the major service centers through the Internet fiber-optic "backbone," will literally explode. The next thing just beyond the horizon is the "Grid."

The Grid takes the web to the next level by creating a computing utility into which users like you and me can simply plug in our own machines, in the same way we plug a toaster into the electrical socket to draw energy. The architecture of the Grid will rely on "peer-to-peer" links that pool together thousands of computers to solve complex problems.

With the "peer-to-peer" concept, we do not need to rely exclusively on selected remote servers to obtain data or computing power. The process now can be spread over many modest machines like ours, linked together to form one giant computing utility.

The first example of such a process was found in a remarkable program developed to analyze signals from radio telescopes in hopes of finding patterns indicative of cosmic life as part of the SETI (search for extraterrestrial intelligence) effort. The software was implemented as a simple screensaver that thousands of individuals could install on their PC. The program would run in the background, periodically calling for a new chunk of unprocessed data and working on it unobtrusively. Three million people have downloaded the program to date, and 600,000 years of PC processing time have been recorded, although no alien intelligence has yet manifested.

The aliens' absence may indicate that radio communication itself is but a transition technology unsuited for cosmic communication. Intelligent races in the universe have probably found something better. Besides, the radio transmissions from your cell phone, not to mention the wireless web, would have been undetectable with the standard radio gear of twenty years ago!

Other programmers have launched similar software to analyze complex molecules and, more generally, to attack problems that had been left to supercomputers. These include drug discovery programs and sophisticated website monitoring tools. The initial uses of the Grid are aimed at the solution of complex scientific and technical problems, but larger applications are just around the corner. There is already a "Global Grid Forum" trying to set standards, and a plethora of companies designing distributed storage networks, load balancers, security mechanisms, and universal file systems that will permit you to reach your files wherever they are in the world, while "caches" will optimize the distribution of information.

The best thing about the Grid is that many of its applications will be free. Based on open systems and widely available public code (developed in Linux or Java), the components of the Grid will permit the first real-life, large-scale implementation of the kind of groupware Doug Engelbart and his colleagues dreamed about many years ago.

The Mesh beyond the Grid

In my own, admittedly fragile model of future networks, the further step beyond the deployment of the Grid is something we could call the "Mesh." In the Mesh, the network fabric extends to more than computers and data servers. It reaches into everyday, ordinary objects. It includes light switches programmed in Java and doorknobs with their own universal resource locator.

Today about one-third of the devices connected to the Internet are not personal computers. They include printers, cameras, special memory devices, and a variety of interfaces with real-world instruments: robots, sensors, and satellites.

This is the kind of technical development that can take the sociologists and the politicians by surprise. It can bring us a new harvest of goodies and horrors, including fine control of energy consumption, vast economic gains, and wonderful environmental benefits combined with nightmarish invasions of privacy at every level.

What kind of applications would be appropriate for the Mesh? One example comes to mind, proposed years ago by Paul Baran but never implemented. In California, where water is a precious commodity, the Mesh could link together all the valves controlling the flow from our lakes and rivers into agricultural irrigation systems. This could save millions of acre-feet of water every year. Equipping these valves and sprinklers with cheap circuits controlled by power transistors is quite feasible, although it may take a gutsy political leader to convince the farm lobby to implement it for the good of the community.

Another example is the creation of information networks using the ordinary power system to distribute news and video to the home while reducing peak energy demands by switching off water heaters and pool pumps at critical times of the day.

All such applications rely on a new generation of analog sensors capable of measuring pressure, flow, temperature, and other physical parameters in order to translate them into digital form and transmit them through the computer network. Using such sensors to help companies with far-flung operations, the Mesh could link together vending machines or remote franchise facilities to signal shortages of goods, distribution glitches, or technical incidents. The banking network of automated teller machines that you find at many urban street corners is an early example of Mesh technology. It can detect physical things like a credit card, a paper jam, or a shortage of dollar bills. In many locations it is also equipped with cameras that can make a photograph of the machine's user.

The fully developed Mesh could transmit images and signals designed to bring into existence a new type of community, spanning the Earth. But the dark side of the Mesh is inherent in its very nature, too. The technology has come a long way from Engelbart's first mouse at SRI and Roberts's first packet network connection at ARPA.

Andy's Shrinking World

"The Internet has changed my life," enthused the young entrepreneur seated next to me at a technology panel. Let's call him "Andy." He went on to explain that he didn't need to leave his house anymore, except if he went on vacation or traveled for business: "I get the news in the morning from a service that screens it according to my stated interests; e-mail keeps me in touch with my business and friends; I order food through the web; I can search for the best deals for all my supplies, and the goods are delivered to my door within the hour. I have vastly expanded control of my universe."

I had exactly the opposite reaction: what a narrow, constrained world this man had created!

Much of my professional life involves use of the web, but my decisions are mostly influenced by external events: chance meetings, unexpected news that catches my eyes in a newspaper left on the counter by a previous diner in a coffee shop, phone calls from faraway friends, or random observations as I shop for two-by-fours at Home Depot or Kodak film at Walgreen's.

The web has become an important part of my family's life—I used e-mail to stay in touch with my children in college a long time before the web even existed—but it does not run it. The world described by Andy seems full of undisclosed biases and restrictions. It filters reality through screens and models designed to optimize his life, but they also impoverish it.

This is the virtual world that collapsed in the opening months of the twenty-first century, the first failed implementation of the Solid State Society.

The collapse of the dot-coms came as a combination of the drop in computer purchases by big firms (they had massively upgraded their networking equipment in anticipation of the Y2K transition) and the realization that valuations of Internet companies by Wall Street were based on pure hype, in the absence of tangible revenue.

It was a collapse of unrealistic promises made by entrepreneurs, investors, and financial analysts who had not understood the nature of the Internet in the first place.

There will be others.

The biases in Andy's world are magnified by the formidable new ability of major enterprises to manipulate what he sees on his screen.

There was a blissful period, just a few years ago (eons in Internet time!), when he may indeed have been exposed to a world of opportunities, with many groups vying for his business and offering their services. But this landscape has closed down around him without his awareness of the changes. The basic institutions on which Andy relies—his bank, his credit card company, his insurer, his travel agent, his broker, his favorite airline—now have access to detailed models of his behavior that can be used to orient his next move in a direction favorable to them.

At a primary level, there is nothing wrong with this. Merchants have always offered special deals to their good customers, and many pride themselves on knowing their habits. We feel flattered when a restaurateur greets us by our first name or when the manager of the local bank comes over to shake hands. We blush with pleasure when a flight attendant discreetly signals to us that we have been upgraded to first class because the computer has determined we had accumulated enough points for special status.

The problem with user profiles on the web is that they do not reflect a particular interest in Andy as an individual. These big companies have similar statistical dossiers about all their customers. They have been trying to influence them for decades, but they had to rely on fairly vague indicators of wealth and behavior: Andy's zip code, his age, what kind of car he drove. The databases could only be correlated in a coarse manner that was often uneconomical—witness the mountains of junk mail catalogs we continue to receive for goods that do not interest us.

Data-mining on the web and the exploitation of cookies have changed all that. Every time Andy orders something, his action is captured and recorded somewhere in a form that can be picked up by advertisers and marketing companies. Firms like DoubleClick can be hired to provide accurate profiles of Andy's economic, political, and personal habits. And if that is not intrusive enough, these institutions will share and correlate what they know among themselves.

Targeting Andy

Suppose my friend Andy decides to refinance his condo. He does this through Bank A, which holds his mortgage. In support of his application, he submits his tax records for the last two years. These identify his employers, his debts, his stock holdings, his salary, and his dependents.

As a happy result of his refinancing, Andy now has a more favorable interest rate and $100,000 in cash. When he looks at his screen over breakfast the next morning, he finds an e-mail message from Broker B offering a very attractive investment opportunity, and a notice from insurance company C for a new, tax-deferred annuity. Credit

Card Company D also mails him a new plastic marvel at a surprisingly low teaser rate.

Andy may be delighted at this sudden flood of opportunities made possible by the web. He may even congratulate himself again for his technical savvy, his ability to use the Internet for such business. In reality, Andy's world is closing around him. Broker B is an affiliate of Bank A. So is insurance company C. They now share the information from his mortgage application. And there is nothing illegal about it.

As a result of all this data sharing, Andy is unlikely to become aware of other investment prospects, possibly offered by unrelated firms. He is subtly being led to consider an annuity that may be a stupid financial move, when a little research in the real world could lead him to better decisions.

A California politician, State Senator Jackie Speier (D-Hillsborough) tried to do something about this by proposing a series of consumer privacy measures in a bill called SB773. It would strengthen a 1999 federal law that tries to safeguard financial information disclosed by consumers on their credit applications.

California generally leads the U.S. (and the U.S. tends to lead the world) in legislation related to the uses of high tech, so it will be interesting to watch what the legislature does with Jackie Speier's proposal.

It looks like an uphill battle.

In response to Bill SB773, banking and insurance interests mounted a major lobbying campaign to defeat or water down its provisions. In the first half of 2001 they spent $7 million on efforts to lobby the California lawmakers, and funneled $500,000 into Governor Gray Davis's coffers, an amount that grew to $10 million by mid-2002, at which time Jackie Speier started running into strong opposition not only from Republicans but from California Democrats from her own party. The financial firms want to be allowed to sell "contact information" such as your address and phone number to outside telemarketing firms, except when expressly prohibited by you.[53]

In other words, you would have to send a specific notice by mail to every institution with which you do business, instructing them not to violate your privacy! And this would not prevent them from sharing your data with other companies if they are affiliates of one another, or if they have entered into a joint marketing agreement. This naturally leaves the door open to every kind of abuse.

In earlier times, the restrictions on information transfer contemplated by the proposed bill may have been effective in giving some measure of protection to consumers. But in the new world of cookies and web bugs, they are grossly ineffective. The databases that are compiled about Andy, about you, or about me are available to anyone who cares to pay for that information.

Andy's impression of an expanding world of opportunities and conveniences is a cruel illusion maintained by major media corporations and financial institutions that are rapidly gaining control of the web.

And of the Grid tomorrow.

And of the Mesh in five or ten years.

There is no evil intent in this trend. Remember, information is control. Andy may believe that he has the best information, but the scope of his awareness is tiny compared to the resources and sophistication of the major databases.

Every communication invention has been hailed as a triumph for original thought and personal convenience, but it is in the nature of mass media to gravitate to the lowest interests of their consumers and to fall under the sway of a reduced number of very large corporations.

There will always be colorful exceptions to the drab flow of commercially supervised drivel. The far corners of the web will remain as a source of innovative or poetic ideas, subversive irritants, and bizarre theories. There will be vast unmined data landscapes ready to be examined and probed by the astute network traveler. But such flowers of intelligence or rebellion may only serve to justify the massive enterprise of data gathering placed at the service of e-commerce by the new architecture of the web. A long time ago the same thing happened to radio and television, two wonderful technologies swallowed up by advertising and drowned in trivia.

At best there will be a point of equilibrium between two forces on the Internet. On one side, we will see the expanding power of major industrial interests like AOL-Time Warner, Comcast, or Microsoft in the U.S., Sony all over the world, Thorn-EMI in Great Britain, Vivendi in France, and Bertelsmann in Germany, trying hard to adapt the web to their traditional mechanisms for market and political domination. On the other side, we will witness the continuing "organic" growth of information structures within the web, amplified by novel forms of viruses and hacking. These new structures, in their chaotic way, will expand on the unpredictability of the old Internet.

Both models are increasingly dangerous for us as free individuals in a sustainable society; both entail further losses of control for their users rather than the harmonious development of human life predicted by the designers of the Internet. And both call for cures that may be worse than the dangers they present.

9

They Want Well-Trained Humans

Once upon a time the Devil decided to visit the Earth, accompanied by one of his young disciples, to see how the tenants were behaving. As they walked along, they came to a town where a big crowd had gathered in the main square, talking excitedly. The Devil sent his companion ahead to assess the situation.

"Master, Master, the news is bad," reported the young man. "These humans have managed to understand part of the Answer!"

The Devil laughed. "Don't worry about it, I will make them believe that they have understood the *whole* answer!"

So it goes with the Internet. The claims made by network advocates about the benefits of the technology for industry and individuals are not wrong. They are simply fragmentary. They gloss over issues of privacy, of government intrusion into everybody's business, of enforcement of arbitrary standards of morality or thought, and ultimately of mind control.

As governments discover the risks associated with specific aspects of network expansion—from electronic fraud to pedophilia to the rise of new antisocial movements—they are under pressure to implement simple-minded solutions that either displace the problem or turn it into a more pervasive and dangerous restriction on ordinary, law-abiding citizens.

As was the case when the SRI-ARC project degenerated into a social experiment with cultist overtones, the web itself could become the tool of a soft, benevolent dictatorship, hardly detectable yet powerful enough to bend the behavior of consumers and citizens to the arbitrary rules of a few institutions.

In some countries, it already does.

An Obsession with Control

The discussion of government's response to the terrorist attacks of September 2001 and its impact on privacy rights is beyond the scope of this book. However, the process illustrates the potential damage that indiscriminate control of communications can inflict on a free society.

When law enforcement agencies demand access to selected e-mail records in order to track down terrorists, not only are such requests legitimate, they are essential to the survival of a free society such as ours. One would even have expected that computer networks would be used to foster greater cooperation among security agencies so that an episode like the terrifying attack of September 11, 2001, could have been prevented.

The danger to civil liberties lies in the abuse of this process in the name of future protection. In the United States most agencies are well aware of the overriding constitutional guarantees under which they must work. Other nations are not so troubled.

Even Great Britain, long regarded as a haven for individual rights and the respect of privacy, has taken a troubling course. As a result of James Ellis's invention of public key cryptography in 1969, British authorities have passed a law compelling users to surrender their encryption keys to the government.

Ellis worked at a place known as GCHQ in Cheltenham, England, a secret communications facility. His work remained officially classified until 1998; but Whitfield Diffie, a Stanford University computer scientist, independently rediscovered the technique with Martin Hellman in 1975 and made it public, to the consternation of government cryptographers. This type of encryption has been used ever since across the Internet, raising all sorts of concerns among governments. In Great Britain, refusing to surrender encryption keys or passwords means two years in jail, and another five if you dare to complain publicly about it, according to a privacy advocate, Caspar Bowden.

Early in 2000 Great Britain also instituted a $30 million system to snoop on Internet e-mail, installing black box recorders on the premises of Internet service providers. As *New Scientist* put it, "The Regulation of Investigatory Powers (RIP) Bill will give security forces unprecedented powers to snoop on Internet users and demand encryption keys."[54]

The proposed technology is not simply offensive. It is obsolete and inept. With the introduction of "always on" broadband communications like DSL, people can set up their own mail servers, bypassing the service providers and government black boxes, where all mail was supposed to be decrypted by the "good guys."

Other countries, such as Singapore, have imposed constraints on the way Internet is used and the information it conveys; but the most extreme measures were taken in Afghanistan on August 25, 2001, shortly before the attack against Manhattan and the Pentagon, when the head of the Taliban government ordered a ban on all access to the Internet.

The decree from Mullah Mohammed Omar, heard on Radio Shariat, is a classic of the genre. It said that the only connection to the network would be "in the office of the Supreme leader, to be accessed by a trusted man."[55] The decree went on to state, "The Office of Communication is instructed to find ways to ensure that use of the Internet becomes impossible. The Ministry for the Promotion of Virtue and Prevention of Vice is charged with monitoring the order and punishing violators."

Toward State-Sponsored Piracy?

Like Great Britain, the European Union has sought to establish restrictions on the confidentiality of communications. Its current version of an international treaty would authorize widespread wiretapping of international communications "to catch criminals." The problem is that the treaty would provide no protection to ordinary law-abiding folks. In the wake of terrorist attacks in September 2001, the move to state-sponsored spying on network users has become irresistible.

Most troubling to Europeans is the absence of open debate about the issues. All deliberations have been closed to the public, prompting an independent user group, the twenty-eight-member Global International Liberty Campaign, to address a strongly worded complaint to the secretary general of the Council of Europe, Walter Schwimmer: "Police agencies and powerful private interests acting outside of the democratic means of accountability have sought to use a closed process to establish rules that will have the effect of binding legislation," stated the complaint.

The treaty could force the opening of corporate networks to law enforcement and make companies liable for misuse of their equipment by hackers or criminals. It could even open the door to government-sponsored snooping into the affairs of another country, and to wider industrial espionage by corporations, an old practice that is becoming increasingly common and dangerous.

Early in September 2001 no less a company than Procter & Gamble admitted that it had hired a specialized security company to steal the secrets of its rival Unilever, and Oracle did the same with companies supporting Microsoft.[56] While most of the snooping was done the conventional way, by going through the trash, access to computer records in the name of security is becoming an easy way to obtain information through an Intranet back door. Indeed, Bill Weber, the executive director of the Society of Competitive Intelligence Professionals based in Alexandria, Virginia, stated that many of the Society's members came from government service, where they would have become familiar with such techniques.

The problem is that absolute computer security is not an achievable goal in these days of global interconnection. Since 1998 a team at Sandia National Laboratories in Albuquerque has been probing top-secret government installations and private sector computers. As of December 2000 this "Red Team" has been successful in infiltrating all the systems it was asked to probe.[57]

Naturally, most firms do not engage in espionage directly. They hire an independent contractor known as a "kite" to provide plausible deniability. So do government agencies. When law-abiding citizens and companies agree to relinquish their privacy in the name of catching terrorists or identifying the bad guys, they may unwittingly provide assistance to another class of even more powerful and pervasive information criminals. In the words of Dr. Brian Gladwell, a former top computer expert at NATO: "If we look at cyberspace we have state-sponsored information piracy."[58] The practice is not limited to the United States. Economic espionage is a game everyone can play. And does.

A Lesson about Security

Twenty years ago an oil company taught me a good lesson about computer security. I never forgot it.

InfoMedia was providing computer conferencing services to a firm that was doing active geological exploration in Alaska. Its teams used our system to coordinate fieldwork with headquarters planners, engineers, and exploration specialists. During a visit to the headquarters of the company, I was invited to have lunch with the managers, and I felt I had to make a little speech about security.

"In the interest of full disclosure," I began, "I should tell you that our system is not providing a high degree of protection to your users. We do encrypt your data by scrambling the bytes around, but an astute hacker could gain access to the file and read it. All we could do is prove criminal intent in court, because the data couldn't get out accidentally. But the network itself is not secure. So I have to ask you, how concerned you are about security?"

They laughed, patted me on the back, thanked me for my honesty, and said: "Don't you worry about us. We are very concerned with security. So concerned that we would never put critical data on your computer; or on our own computer, for that matter. The guys who are spying on us wouldn't be interested in hacking into the database on your system. The information is too old for them by the time it's reduced to a database."

"What are they looking for, then?"

"They want the rocks themselves. As they come out of the ground. The CIA is a bunch of amateurs compared to those guys. They buy our geologists."

Another manager broke in: "I'll tell you how concerned we are with security. The last time we negotiated a joint bid on a North Slope property with another company, we leased a train and we put both of our teams on it with the secretaries and the accountants, and we ran that train in circles in the Canadian wilderness for three days, until the bid was typed up, signed, and sealed."

I was very relieved by that answer, which came from the real world. Ever since, I have looked at the issues of computer security, and the reliance on encryption, with a more cautious and skeptical eye.

The Special Case of China

There is an ongoing dispute between China and the United States about the Internet, and it could flare up into a full-scale net war some day. On December 5, 2000, a top Chinese official, information minister Wu Jichuan, directly accused the United States of "cultural, economic and technical hegemony of the Internet."[59] Much of the argument had to do with unfair charges. A majority of Internet traffic between developing countries is routed through the U.S., which controls the cables, charging Asian economies an estimated $5 billion a year. But the real dispute had to do with control of content, including Chinese-language domain names.

Responding for the U.S., Gregory Rohde, assistant secretary of commerce for communications and information, "dismissed most of the criticism as backwards-looking assertions of national sovereignty that reflect fundamental misunderstandings about the Internet's development."

The U.S. seemed to be saying that the Chinese were sadly out of touch with the realities of modern computer technology. Rohde added that disagreements over domain names should be left to nongovernmental entities like the privately run ICANN, the Internet Corporation for Assigned Names and Numbers, whose members are elected.

The Chinese smiled under the insult and politely pointed out that ICANN was created by, and reported to, the United States government. Wu went on to say that Beijing planned to restrict the use of bulletin boards, adding, "Don't misinterpret this. The Chinese government absolutely is not saying people can't use these things, but we must find a more healthy way to manage them to ensure the protection of individuals' reputation and privacy."

He acknowledged that such controls would be hard to enforce. But the Chinese government is not giving up on its repressive efforts. Nineteen people have already been arrested for "spreading subversive information" via the Internet. Under its campaign for "spiritual cleaning," the Ministry of Culture has closed down some eight thousand cybercafes throughout China.[60]

On November 14, 2001, Wang Jinbo, a twenty-nine-year-old Internet user, was brought before a court in Junan district for "inciting to subversion of state power." He had written e-mail messages calling for rehabilitation of the student movement that demonstrated on Tiananmen Square in 1989.

In its efforts to counteract Chinese censorship of the Internet, the U.S. government has plans to facilitate anonymous access to websites everywhere. This creates a very interesting paradox: while Washington is frowning on strong privacy within the United States, it is actually encouraging it in other parts of the world.

In connection with the International Broadcasting Bureau, the parent company of the Voice of America, a California company called Safeweb that specializes in Internet privacy is developing a computer network that would allow Chinese citizens to get around their government's censorship of the Internet.

Many sites, including those of the Falun Gong movement, Amnesty International, and the Western media, are censored by Beijing. The

Safeweb software, known as "Triangle Boy," would allow Chinese users to reach its servers on an anonymous basis. While censors could block a few of the programs, they would presumably be helpless when thousands of them were activated. Since there were an estimated 26 million Internet users in China in July 2001 (versus only 9 million at the end of 1999), the censors' task would presumably become unmanageable.[61] Remarkably, at a time when the U.S. government is cracking down on anonymous usage of the Internet and spying on e-mail through software like Carnivore, Safeweb is a California start-up funded in part by In-Q-Tel, the venture capital arm of the Central Intelligence Agency.

Privacy under Attack

Privacy is a funny thing, whether you are Chinese, French, British, or American. Over breakfast in sunny Woodside, California, I asked Brian Pinkerton, the young vice president of engineering for a major Internet service provider, what he thought of the current trends in the sharing of user information.

"It's a strange contradiction," he answered. "If you ask users to rank-order the factors they really care about in a web service, they will always put protection of privacy near the top of the list. But if you offer them a $25 rebate on a subscription to some silly magazine or their next trip to Disneyland, they will sign away all the rights to their personal data."

Both of us came from a design philosophy that placed users at the center of the information process, but the new systems treat them as consumers with a reduced set of rights and little access to their own information. They also make it easy to spy on your neighbors. Witness the case of one James Jackson of Memphis, who selected prominent people from *Who's Who in America* and found no difficulty in buying dossiers on them from so-called information brokers on the Internet. The dossiers included Social Security numbers and locations of bank accounts.

A San Francisco federal judge, Alex Kozinski, discovered how little privacy was left on the web in the spring of 2001 when he became aware that a monitoring program screened everything he and his colleagues were doing over the network.

Judge Kozinski is no helpless computer user. He works with the United States Court of Appeals for the Ninth Circuit, based in San Francisco. It is the largest of the nation's twelve regional circuits, covering nine states and two territories. Upset at the discovery of the monitoring software, a council of the circuit's appeals and district judges ordered their technology staff to disconnect the monitoring program while they reviewed the situation with the Administrative Office of the Courts (AOC), a Washington bureaucracy.

Judge Kozinski distributed an eighteen-page memo arguing that the monitoring was a violation of antiwiretap statutes.

In response to this uprising, the AOC manager, Leonidas Ralph Mecham, explained that the software was necessary to enhance security and reduce computer use unrelated to judicial work, adding that a survey by his office had "revealed that as much as 3 to 7 percent of the judicial browser's traffic consists of streaming media such as radio and video broadcasts, which are unrelated to official business."[62]

Chief Judge Mary Schroeder of the Ninth Circuit responded that the concerns were overblown, adding that the policy had nothing to do with the overloading of the system but with Mecham's preoccupation with "content detection." She added, "We are concerned about the propriety and even the legality of monitoring Internet usage," implying that the software may have violated the Electronic Communications Privacy Act of 1986, imposing civil and criminal liability on any person who intentionally intercepts "any wire, oral or electronic communication."

Commenting on the situation, the *New York Times* quoted professor Jeffrey Rosen of the George Washington University Law School, author of *The Unwanted Gaze*:[63] "This drama with the judges reminds us of how thin the privacy protections are." And the article went on to remind readers that 63 percent of American companies already monitored employees' computer use.

Who Is Kidding?

It is encouraging to find federal judges suddenly upset at the creeping use of computer technology for spying on users' personal habits, but many people in the electronics communication field have long taken it for granted that such practices were common. Police installed the first telephone wiretap in the United States in the 1890s, and law enforcement agencies routinely get court orders to spy on electronic mail, file transfers, instant messaging, and bulletin boards or chat rooms.

In his book *The Rise of the Computer State*,[64] written well before the Internet age, David Burnham had already observed that "computers in recent years have in fact leapt the boundary that the critics argue separates man from machine." Such important inventions as speech recognition or language understanding, whose early development we reviewed in the first part of this book, have opened the way to abuse: "The computer may lead to more wiretapping and bugging by reducing the economic barriers to eavesdropping for agencies and organizations that can afford to purchase voice recognition systems. But the computer will not only knock down the obstacles that tend to protect the words we speak. More importantly, it someday may increase the vulnerability of the thoughts we think."

In chapter 2 ("The Digital Society") we looked at applications of computers to law enforcement, with the observation that such tools do more to change the nature and patterns of crime than to eliminate it.

Over thirty years ago Congress defined the rules for intercepting telephone calls. Title III of the 1968 Omnibus Crime Control and Safe Streets Act authorized law enforcement to circumvent individual rights to privacy, provided that a court order was obtained.

In 1986 the Electronic Communications Privacy Act (ECPA) extended these provisions to e-mail and data processing. But the content of messages was still off-limits. Law enforcement could merely record originating phone numbers and traffic data.

All those nice legal safeguards went out the window in the late 1990s.

The advent of the web has enormously increased the ability of governments to spy on their own citizens, not only in voice recognition but in all other forms of communications. The FBI was quick in recognizing this potential. It developed a system called Carnivore, ostensibly to capture e-mails exchanged by "bad guys" under a strict procedure of authorization by federal courts. In reality, Carnivore does not necessarily discriminate between good guys and bad guys as it intercepts e-mail traffic flowing through Internet service providers (ISPs), who are forced to comply. Only one did try to resist federal pressure: Earthlink of Atlanta was the only ISP to challenge the right of the FBI to install Carnivore on its machines.

It lost the case.

As Dan Hester wrote in *TechWeek*, an industry magazine:[65] "Federal gumshoes can now invade your privacy by filtering your e-mail, in effect entering your life without kicking down physical doors." Which opens the probability of other abuses since the operators of Carnivore can target their victims without the benefit of a search warrant or the supervision of a judge.

When the system was exposed, there was a national uproar against such practices. The government's response was to avoid the issue. In July 2001 Attorney General John Ashcroft visited Silicon Valley Internet firms. After a meeting at Verisign, he announced a series of measures targeting cybercrime, beefing up ten special units called "CHIP," for computer hacking and intellectual property. But journalists reported that he dodged the question of his own services' abuses of individual privacy rights on the Net, assuring his audience that the FBI no longer had a system called Carnivore.[66] It turned out only that the system had been renamed "DCS1000." How and when DCS1000 gets used remains subject to wiretap laws and strict rules of evidence. But history has shown that abuses were inevitable.

Orwell Was an Optimist

The National Security Agency (NSA) had been doing the same thing as the FBI for decades, with a far wider system called "Echelon." Aided by the intelligence services of several other countries, Echelon intercepts virtually every kind of electronic communication from satellite to microwave, telephone, and Internet messaging.

When the existence of Echelon was publicized in the European press in 2000, many voices were raised to condemn the system as another example of extreme American bullying. The protests were not very credible: European countries had long shared in the "product" of Echelon, in particular the French, who were eager to keep track of Arab terrorists. European countries have been operating similar systems for years, keeping tabs on their own citizens.

Everybody can play this game, and the public protests themselves are hypocritical.

In the words of one of my friends, an expert on Internet security, "Orwell was too optimistic. There isn't going to be a Big Brother. In the world to come we will all be each other's Big Brothers."

Even earlier, French sociologist and World War II Resistance leader Jacques Ellul had observed that:

> The techniques of the police, which are developing at an extremely rapid tempo, have as their necessary end the transformation of the whole nation into a concentration camp. This is no perverse decision on the part of some party or government. To be sure of apprehending the criminal, it is necessary that everyone be supervised.[67]

The obvious answer to such criticism is that law-abiding citizens should have nothing to fear from government surveillance. If you do nothing wrong, why should you be concerned if a policeman scans your e-mail, or a counterintelligence agent reviews your purchases from foreign countries? Come to think of it, why shouldn't every child be fingerprinted? Why shouldn't our DNA be recorded in a government databank? Advocates of such measures frequently point out that this would facilitate fast reaction to kidnappings, and protection of innocent victims.

The problem lies with the vesting of power in unseen agencies that are not amenable to control by individual citizens, or are only in extraordinary circumstances—when the victims happen to be federal magistrates, for example.

The power to review our files and messages ends up in the hands of civil servants, most of whom are honest, hardworking members of the community. Others may not be so pure in their activities. The FBI and the NSA have been known to harbor foreign spies and criminals. Remember Mr. Robert Philip Hanssen, who for years, as head of FBI counterintelligence, leaked American secrets to Russia? Or his counterpart at the CIA, Mr. Aldrich Ames? And what about the frequent revelations that federal intelligence agencies have been infiltrated by sects with extremely bizarre beliefs and an allegiance to their own hierarchies?

Beyond these obvious concerns, why should we rest comfortably while modern governments train hundreds of their agents as professional hackers, who will find employ in dozens of private investigation companies when they retire or leave the public sector? Many of them will go into legitimate security firms, but what about the others? What does a forty-five-year-old intelligence agent trained in network penetration do when he or she becomes a consultant? Who hires such skilled people? And for what kind of work?

A Prologue to a Farce or Tragedy

The real answers to the questions we just posed—questions that occur with increasing frequency among company CEOs and their boards as business relies more and more on the use of the Internet—are kept secret. Even Congress has a hard time finding people to testify truthfully on these issues, clouded as they are by technical jargon and laws protecting classified information.

Sometimes the secrecy is justified. There are nasty problems in the world, as Bin Laden and his disciples have proven all too clearly. Any progress in tracking down terrorists, or the manufacturers of biological and chemical agents, obviously is welcome. But trained secret agents have ways of exchanging information that are not only difficult to break but nearly impossible to detect. You can hide a message in subtle changes of color in selected pixels of a digital photograph, for example, using open-space "steganography" software. These techniques are known to most programmers, amateur cryptographers, readers of Tom Clancy, and members of Al Qaeda.

The reality of the situation is that the secrecy is a sign of continuing government encroachment on the thoughts of citizens; and more generally, of increasing encroachment by very large media companies with even more pressing needs than modern governments.

This fact represents a long-term threat to the fledgling Internet industry. As a venture capitalist who invests in high tech, I have to worry that the web will be perceived as an increasingly corrupt police state overlying a maze of dark alleys and unsafe practices outside the rule of law. The public and many corporations will be reluctant to embrace a technology fraught with such problems. The Internet economy will continue to grow, but it will do so at a much slower pace than forecast by industry analysts.

As a result of such concerns, politicians are already under pressure from a business community desperate to reassure its own customers that shopping and talking on the web is safe and fun.

The problem is that major network interests are reluctant to relinquish their hold on consumer data, so the new laws may remain ineffective. Ordinary users of the web have no way to detect, or even to comprehend, how the massive datamining companies track their behavior and influence their use of the Internet. In frustration, Deb Hooper, an Internet privacy expert, put it clearly: "I don't want the government writing privacy laws they can't or won't enforce."

What can we do about it? The first step is to safeguard our own information: "The best way would be for all of us to boycott sites that don't offer opt-in (the opportunity to prevent a company from sharing information about us)," says Deb Hooper.

In an essay against government attempts to restrict scientists, Dr. Edward Teller, who can hardly be accused of ultraliberal leanings, once reminded his audience that:

> Secrecy is not compatible with science, but it is even less compatible with democratic procedure. Two hundred years ago, James Madison said, "A popular government without popular information, or the means of acquiring it, is but a prologue to a farce or tragedy or perhaps both."

Madison could hardly have anticipated the web, the Grid, and the Mesh. Yet his prophetic words could be aptly extended to the Internet today, and to the even more sophisticated forms it will take in the future.

Part Four

How We Can Save the Dream

The gradual erosion of privacy is not just the unimportant imaginings of fastidious liberals. Rather, the loss of privacy is a key symptom of one of the fundamental social problems of our age: the growing power of large public and private institutions in relation to the individual citizen.

—Walter Kronkite in the foreword to *The Rise of the Computer State* by David Burnham, 1984

10

Four Essential Principles

We have some decisions to make. The Internet is here and its impact is growing. So far it has been felt mostly in the business world, where it has greatly improved productivity. It has allowed firms to restructure order processing, financial transactions, customer relationships, internal business rules, procurement, delivery tracking, and inventory management throughout far-flung operations.

As individuals, most of us have seen the Internet only as a simple messaging channel ("You've got mail!") or as a convenient way to buy books, look up old flames, and research family roots.

Tomorrow will be quite different.

The Internet will affect us at three levels: as citizens, as members of communities, and as private individuals. On all three levels it is urgent that we educate and protect ourselves, and it is imperative that we become involved.

"How could we possibly be involved?" most people will ask in some consternation. The technology is complex, obscure, intimidating. There is no regulation around to help enforce our most elementary rights. And what are our rights in this new medium, anyway? They get redefined at the whim of corporations large and small. When AOL swallowed up Netscape, the two companies merged their user lists. This wiped out the account name I had been using, and I was never told that someone else had been using the same name on another system. When AT&T dropped the "At-Home" service of Excite, they did not bother to provide a backup for their hundreds of thousands of users, whose entire e-mail history vanished into the black holes of cyberspace.

The sheer size and power of Internet juggernauts seem to preclude any effective action from mere users.

The Seven Brothers of Electronic Media

If "Seven Sisters" used to dominate the oil business, seven brothers now rule over electronic media. They are: AOL Time Warner, Vivendi Universal, Viacom, News Corp, Disney, Bertelsmann, and Sony. Among these, AOL and Vivendi have been the most aggressive on the web. As an article in *The Economist*[68] points out, "AOL, after all, has 34 million subscribers whose eyes tend to fall on whatever content is parked on the service provider's home page."

For these huge conglomerates, vertical integration is a tempting business model. If one of your studios is ready to release a mediocre flick or a song of dubious appeal, it is very convenient to push it to the top of the charts by preselling it to your captive Internet subscribers, flooding the market with teasers and good news, and pushing the derivative products through your other channels: radio stations, theme parks, book publishers, newspapers.

The consequences for us as customers are also clear. Not only are we fed a steady diet of bad entertainment and biased news, but we miss a chance of even finding out about alternative products.

Fortunately, there are indeed solutions we can apply at all three levels to save the dream. And the conglomerates are less powerful than they want us to believe.

As users, we should take comfort in one simple fact: on a system of computer networks that can potentially link 600 million users, no single institution or company is going to call the shots for very long. We can reclaim the ultimate power, if only we educate ourselves in the best ways to apply it.

Of Monopolies, Dark Matter, and Chastity Belts

Once you have experienced the convenience and the thrill of web interaction, you can never go back. You catch the spirit of the Internet and you keep it, at work and at home, to enrich your life—also enriching Intel, AOL, Yahoo!, Microsoft, and others in the process. But that all happened yesterday. We now enter a phase where the development of the web will be severely threatened unless it is protected from encroachment by business, abuse by government, and intrusion by criminals.

Who would suffer from a slowdown in web development? All of us would, starting with business itself, constrained in its expansion and in productivity gains; government would find itself increasingly sclerosed, befuddled by information structures of overwhelming complexity, vulnerable to the very weapons it is deploying to defend itself; and ordinary people would see their world shrink and wither, unable to deliver the promise of a viable information world.

This slowdown is already evident in the many deficiencies present on today's Internet. At least 5 percent of the network is unavailable to you at any given time, hidden by unreliable routing or simply misconfigured. This mass of dark matter and unreachable data is likely to grow. And vertical consolidation of big media companies tends to restrict users' horizons even more.

Everybody loses in this game, because the overall utility of the Internet simply decreases as these firms attempt to turn their subscribers into captive customers. The same goes for banks and credit card companies that distribute your personal data to their "affiliates." In the long run, such practices actually hurt those who try to enforce them.

I am old enough to remember the early days of networking, when international data traffic became available from most Western countries. I used to travel to Europe with a portable terminal capable of dialing into ordinary phone lines to access U.S. computers. The German telephone monopoly, the DeustcheBundesPost, saw this new practice with alarm.

They forced users like me to lease a special modem from them at a high price. This was all the more ridiculous in that I might be in Berlin or Munich only for a few days or a few hours, with no time to go to the post office, fill out complicated forms, and wait for a shoebox-size modem to be delivered. Furthermore, I carried a new kind of portable terminal that was equipped with rubber cups (we called them "rabbit ears") that allowed me to use any ordinary telephone in a hotel room or a home and plug it into the system. The cups contained an acoustic coupler that did the same job as the post office's expensive shoebox. Never mind, we were told. We had to lease the modem anyway, even if it was destined to remain on a bedside table, unused, while the rabbit ears did the job!

Such stupid business practices find their parallel today in the efforts of Hollywood companies that are lobbying legislators (notably

Senator Dianne Feinstein of California) to put physical locks on CD and DVD readers in futile efforts to prevent piracy of music and video.

These companies, like the old DeutscheBundesPost, do have a legitimate right to control their products and their services, but trying to impose a new kind of chastity belt on every computer's disk drive is not the way to do it. Any technological explosion flows to the ultimate advantage of the media giants, piracy or no piracy. It is the nature of their entire business that is shifting under their feet, like the massive motion of the Earth plates that will some day bring Los Angeles to the same latitude as San Francisco. The book business, which used to view the young Internet as a diabolical monster, now thrives in the environment of Amazon.com and thousands of online bookselling websites. But Hollywood has yet to learn that lesson.

Four Principles for a Healthy Web Future

Nobody has written a simple Charter of the Rights of Internet Users. I do not presume to start writing such a document here, but only to list the four most basic principles that seem obvious to me as I summarize the earlier observations of this book. From these four principles we will draw practical guidelines we can start applying in everyday usage of the web.

THE FIRST PRINCIPLE IS FREEDOM OF ACCESS

There should be no restriction on where you can point your browser, provided you adhere to the rules of every particular site you visit. Any country or media service that places constraints on access not only will deprive individuals of a basic opportunity but also will suffer drastic economic, social, and political consequences as its culture becomes impoverished. There should be no block placed on legal uses of the web.

THE SECOND PRINCIPLE IS PROTECTION OF PRIVACY

There are circumstances when we, as individual users, will voluntarily loosen or relinquish our control of privacy (we already do when we allow web services to track our use of the web, or when the phone company keeps a log of places we've called), but we should always know what information is collected, where it goes, and what is done with it. This, too, is ultimately in the interest of business, because it makes fraud more difficult to perpetrate. Such rules are already included in European privacy regulations, and they should be extended everywhere.

If privacy is not protected, our citizens will gradually lose trust in their institutions and will stop exercising the precious rights that define us as members of the free world and as a democracy. And this, in turn, could kill the Net.

THE THIRD PRINCIPLE IS ASSURANCE OF NEUTRALITY

The web, as a transport mechanism for knowledge, should never bias the information it gives you. If a search engine returns references based on how much it has been paid to lead you in a particular direction, as most of them do today, how can you ever trust the Internet to do unbiased research? You are simply reading ads when you believe you are retrieving information. This is bad for everybody, but especially for the web infrastructure itself, in the long run. As the infrastructure deteriorates in the range and trust of the data it carries, the entire network becomes compromised and vulnerable.

THE FOURTH PRINCIPLE IS STRUCTURAL INTEGRITY

Corporations have begun taking control of the inner structure of the Internet. The packet-switched concepts are being compromised by virtual private networks, "tunnels," and priority channels among clusters of storage devices acting as caches or as vast reservoirs for corporate information. This internal nervous system of the web, in turn, is increasingly controlled by vertically integrated firms.

This narrows the range of what you can see, what you can get, what you can buy, even the ads you can read. It blurs your awareness of the information world. It biases the integrity of the transport mechanism itself. In the end, it strangles competition around the web, and again, decreases its usefulness.

If this were allowed to build up, tremendous impoverishment of the web would result, similar to the state English author Colin Bennett described in reference to a mind where all anomalous noise is filtered out: the entity we call the Internet "would become little more than a car park papered over with ancient copies of *TV Guide*. When that happens, the living are indistinguishable from the vast hosts of the dead."

Facing the Consequences

Where do the four principles take us?

- As private individuals, we have the right to protect our privacy and to resist encroachment from greedy corporations and snooping bureaucrats. We must take elementary precautions to secure our data and restrict the information we make available for data-mining by advertisers. These precautions (or "countermeasures") deserve their own section. We will review them in the next chapter.

- No one should be able to restrict our effective legal use, as community members, of the web itself, in pursuit of our own professional, spiritual, and creative goals. The Internet is a wonderful tool to gain information and to motivate others to act in concert with us. We should be able to use it to associate freely, and this leads to very creative initiatives. Some U.S. cities, as we will see, are already doing this successfully.

- As citizens of a free country, we should be protected against the biases of commercial interests. We should be encouraged to inform political leaders of the abuses we observe, and allowed to build alternative ways to access the information we want. We should be free to use the Internet itself to remain informed on threats to our freedom and to influence legislation on key items.

Accordingly, in the rest of this chapter I will attempt to outline specific actions we can already take.

Actions We Should Take as Citizens

Most state and federal legislators, as well as judges and public regulators, are as ignorant of the issues that surround the Internet as most of the citizens they are supposed to represent and protect.

Worse, they are courted by lobbyists whose interests are antagonistic to the protection of common users. A lot of money is at stake. The insurance and banking industries, the communication conglomerates, the entertainment giants, all would prefer to continue making the rules and reaping the benefits from their version of the Internet, which they see only as a convenient conduit to their current services. They want to put a lock on every disk drive and track every transaction. They will resist change as fiercely as the music studios or Hollywood interests fighting for the enforcement of obsolete business principles.

This raises a huge issue that goes far beyond the appropriate uses of technology.

As my publisher, Frank DeMarco, has reminded me, the American system of government is unique in that it anticipates a series of checks and balances to preserve social peace and personal freedom within a remarkably dynamic economic construct. It would be ironic if the Internet, built on principles of organic growth and free enterprise, were allowed to become a threat to these basic American values.

Congress is currently under enormous pressure to pass Internet legislation that will do away with personal communication privacy, authorize large-scale monitoring of e-mail content, and render computer users powerless in the face of massive data-mining of their medical, financial, and political information—all in the name of protection against the threat of terrorism. This is as absurd as saying that we should do away with our freedoms to make absolutely sure no enemy can abuse these freedoms. This trend, if it continues, will actually weaken our society and facilitate the work of those who would use the Internet as a tool to attack it.

Even in the pre-Watergate era, Senator Samuel J. Ervin Jr. observed, "Once people start fearing the government, once they think they are under surveillance by government, whether they are or not, they are likely to refrain from exercising the great rights that are incorporated in the First Amendment to make their minds and spirits free. . . . And in the long run the government is going to suffer from the effects of this as much as the citizens are to suffer the loss of their freedoms."[69] His words are even more applicable today.

If you are concerned with issues of security and privacy, the first important action is to keep yourself current on these issues and to make your voice heard at the state and federal level. To that effect, there are some important people and organizations you can get in touch with, and some websites that track relevant information.

Specific Resources

Privacy should be the first concern for all of us. In this respect, Representative Dick Armey, a longtime critic of the FBI's Carnivore system, has tried to emphasize oversight of surveillance software. Note that Carnivore itself is being superseded by a new architecture that would concentrate Internet traffic in key locations where all packets, and not just e-mail, can be wiretapped.[70] The Electronic Frontier Foundation (www.eff.org) and the Electronic Privacy Information Center (www.epic.org) provide solid privacy information on the web. Marc Rotenberg is the executive director and David Sobel is general counsel of EPIC. The Privacy Foundation maintains its own website at www.privacyfoundation.org.

Among other important resources: Cory Doctorow is outreach coordinator for the Electronic Frontier Foundation in San Francisco. Another useful site is www.slashdot.com, which keeps an up-to-date notice on "Your Rights on the Net" with pointers to the latest information or threat.

Internationally, the most active group, to my current knowledge, is the Foundation for Information Policy Research (FIPR), based in London. It was created in 1998 by Caspar Bowden and Ross Andersen. Also in London is the civil rights group Privacy International, whose spokesperson is Simon Davies.

Internews, presided over by David Hoffman, is an international nonprofit organization supporting open media.

The Center for Democracy and Technology, an Internet policy organization, is directed by Jerry Berman. Hoffman and Berman cochair the Global Internet Policy Initiative. All these organizations also work to prevent major industrial conglomerates from getting a stranglehold on digital technology.

Creative Actions We Should Take as Community Members

As the larger media companies like AOL and Sony are becoming vertically integrated in the same way as European conglomerates like Vivendi or Bertelsmann, which own book publishers, phone companies, newspapers, ad agencies, radio stations, movie studios, and even the distribution pipelines for their products, Congress and the FCC should be protecting us from any abuses. But the current relaxing of media rules in Washington is only encouraging the emergence of huge information and media monopolies.

A single current example will illustrate how apathetic consumers can be hurt in the current war for control of the Internet. The same Hollywood interests that lost the fight to ban the VCR in 1982 are now trying to establish and enforce a mandatory standard for every device and technology using digital TV signals. As remarked by Cory Doctorow, this would "give them a veto over all digital television devices, including those that interface with personal computers." The Electronic Frontier Foundation is trying to alert legislators to this threat. This is the kind of issue that is hard to place squarely before legislators who have little or no background in the arcane world of networking.

Technology marches on, fortunately, and two developments can play a big role to free up communities from the ever present threat of encroachment, censorship, or outright control by commercial interests. They are municipal networks and personal Internet servers.

Municipal Networks

For a relatively small up-front investment, your town can put up its own local network and allow all its citizens to connect to it at a modest rate, say twenty dollars a month. Such a network can provide connectivity at a bandwidth of 12 megabits per second, much faster than what passes today for "broadband" among telephone and cable companies that are trying to control residential usage.

The technology used can be either fiber-optic or wireless transmission.

Just as some cities (such as Palo Alto, in California) manage their own electrical generators, which often permit them to avoid blackouts when the power grid becomes unreliable or corrupted by greed, your town can force competition with cable companies and Internet service providers.

A prime example of such a successful implementation is Ashland, Oregon. The experience of Ashland (described at their website www.ashlandfiber.net) makes it clear the network is not simply an alternative to traditional cable TV service: "The fiber optic rings that weave through the city's neighborhoods are unlike any other system in Oregon. It's so revolutionary, software companies are scrambling to create state-of-the art products to make use of our network's capabilities: a network that will provide us with incredible speed, unquestioned reliability, competitive pricing . . . and unlimited possibilities."

As my friend Brian Pinkerton observes, "The less the content companies can own the last mile, the better!" Such municipal systems can provide the "IP dial-tone" better than the phone company, as well as elementary services like e-mail, and they will let you manage your connection to any Internet service you want. They will give you back the control of which information you want to see, and when.

In Palo Alto, California, two groups are already working to establish radio networks: PAFree.net has a citywide focus, while a co-op team in College Terrace is flooding its neighborhood with a grassroots WiFi network (Collegeterrace.net) for legal sharing of Internet resources.

Personal Internet Servers

The second development on the horizon could turn every home into a fountain of information, music, and transactions for the World Wide Web, and even a source of movies and reportage on local events or on big issues, bypassing the control of official studios. This new form of popular expression across the entire network can defeat centralized media obsessed with fighting the dissemination of free information.

At a time when over 65 percent of U.S. households own a PC, of which 93 percent have an Internet connection, the linkages among interest groups represent the major untapped potential of network technology. Examples are unlimited: In today's world kids interested in the latest videogame or in a new educational program; seniors eager to discuss travel, bridge, or genealogy; individual writers publishing a weblog; groups with a fascination for art, music, language, politics, philosophy, or architecture—all have to find their way through a bewildering maze of bulletin boards, e-mail services, and websites—each with its own set of passwords and complicated subscription and security requirements.

Tomorrow these groups will be able to run their own servers at home, and to host any personal information they want to make available to the world at large. It is at that point, in my opinion, that the full power of the Internet architecture will be realized.

If you were a member of a community group such as a church, a school, a parent-teacher organization, a support structure for a presidential candidate, a historical society, a Madonna fan club, or a scientific team tracking sunspots, how would you go about installing your own server on the Net?

Today this would involve significant costs and skills. You could start with a modern PC, but if traffic expands you may need a more powerful computer such as a high-end HP, IBM, or Sun machine running under Unix or Linux. This would require someone with systems expertise to maintain it. You would connect it to the Internet through a "fat pipe" such as a T1 line, since your cable modem or DSL telephone line could not carry enough traffic.

Although this is quite feasible for a medium-size business, the project represents too high an investment for most community groups today. The computer itself may cost less than $5,000, but the connectivity will run to about $1,000 per month, plus the salary of a part-time systems technician. And this would not cover the maintenance of the site content—the information you want to disseminate to the outside world.

For this reason most groups are content to have their "web presence" hosted by a specialized company or their communications vendor.

There is no reason for these limitations to remain in force in future years, especially if you live in a town that offers its own independent network. With the acceleration of demand, the price reduction of hardware and the explosion in bandwidth, we have to be prepared to take advantage of new opportunities to make our voices heard. We can imagine not only community groups but also creative individuals running their own video broadcasting services, their database access systems, web diaries, even artistic and news programs, in a completely independent way.

The Next Phase of Web Development

What are the consequences of these trends? The reliance on obsolete business structures, on mega-studios and media giants, on communications monopolies, can be broken or bypassed. These big companies will continue to flourish, providing much of the content, but their business model will be altered. More important, they will have lost the ability to control the behavior of their users, to censor or bias content, and to place a stranglehold on the amount of information individuals wish to exchange with one another across the globe.

This is only one vision of what a truly useful, productive Internet future might hold. In this scenario, the elements of the Net would be a large number of independently operated servers linked through community-owned, regional services like the Ashland fiber loop. There are probably many other ways to arrive at a similar result.

Naturally, this vision will not arise without a fight. Conflicting forces will always try to reassert control over the immensely attractive marketplace of the Internet. But the network itself is our best weapon to detect such threats and find creative responses. It will be a battle worth watching, a real war for the very spirit of that amazing invention we call the Internet.

11

Your Personal Countermeasures

In the previous chapter we spoke about what could be done on the national and community level. But what about the most obvious threats to our privacy and security, those that are all around us on the Internet every day? In this area, too, you have important weapons as an individual, and you should be using them actively.

First, a reminder of some basic facts. If you have a broadband connection to the Internet such as a DSL line, a cable modem, or a satellite link, you may have been given a fixed "IP address," the Internet equivalent of a permanent phone number. That means that hackers and other potential intruders can always find you at the same place. In that situation, you must at the minimum install a firewall program to protect you from most attacks.

If you use a dial-up line, or a basic DSL connection, then you are not permanently hooked up to the Internet, and you do not have a permanent IP address. It will be more difficult, but certainly not impossible, for others to direct intrusive software at you, but you still require some level of protection. Sophisticated hackers, some of whom are trained professionals, use "port scanners" to find out who is online and attempt penetration.

Most of us have two major reasons for needing protection from intrusion, the first obvious one being that we would feel violated if our private data were lifted from our files or tampered with. But the other, equally important reason is that viruses can be extremely destructive. Whether you are a researcher, a businessman, or a writer, they can wipe out days, months, or years of your work. They can even put you out of business. Recovery, when it can be achieved, may take long hours of hard and tedious reconstruction from backup files. And who has time to keep complete backup files anyway?

Whatever system you use to access the Internet, you can assess your level of vulnerability by taking a free online test offered by Steve Gibson (at www.grc.com), which is called "ShieldsUp!" It will tell you how open your computer is, as seen from the web, and how attackers might attempt to compromise your privacy.

Who Are the Attackers?

The real answer is: nobody knows. Who might want to intrude on your activities or mine, and why, is an open question. The media are always quick to answer, "The hackers are guilty!"—by which they mean those ugly nerds who spend their time trying to disrupt the work of normal people and well-behaved corporations. The typical hacker, we are told, is a solitary teenager who can't get a date and focuses all his unhealthy energies on the compilation of destructive programs and viruses because he derives a sense of power from such activities, or hopes to get even with what he perceives as an unjust world.

This "typical hacker" may have existed at one time, but he has long vanished from the scene, along with the lonely cowboy and the prostitute with the heart of gold.

Once in a while the FBI does catch some unfortunate high school kid who accessed a government computer by guessing the right password. (Hint: the most common password is "system," "admin," or the name of a dog.) We watch on television as a team of heavies in riot gear surrounds the house, blows down the door, courageously wrestles the grandmother to the floor, and emerges triumphantly with the kid in tow and armfuls of diskettes.

I once caught such a young hacker on my company's conferencing system (actually, he caught me) and I sent him a message asking how long he had been writing programs.

"Four years," he wrote back. He had broken into another system where the operator, duly impressed, had given him a free account.

"How old are you now?"

"Fifteen," came the answer.

Talent like that should not go to waste, and the kid definitely doesn't belong in a jail where he can share his insights with a few hundred hardened criminals.

Real hackers, who are increasingly older, sophisticated, and occasionally female, contrary to their media image, are probably not very interested in your little secret affairs or the archives of your transactions with Amazon or your bank. What they look for is raw computing power. If they can harness your computer, and a few thousand others like it, they can use them to mount an attack of a really big target, like eBay or Yahoo!, by trying to overwhelm it with a barrage of messages. Or they can use it to hide their identity as they take control of some really interesting process on the web. Most likely, they will use your machine, along with many others, in a massive search for a site that contains the ultimate hacker's booty: a database of credit card numbers.

These people are not interested in your Visa card. They want the central file at Visa.

This is illegal and terribly dangerous, of course, but it only involves you as an innocent bystander. You should protect yourself against being used that way, but there may be little permanent damage if you are attacked. Most likely, you will never be aware that your computer has been used that way.

The real criminals on the Internet are no longer hackers. They are professionals. They work inside the firewalls. Many of them have been trained by the government. They may be former agents of a "three-letter agency" who have become consultants for some private firm. Or they may have gone over to the "other side" and joined a criminal group. Those are the people who will try to steal your identity, look up your bank records, or simply gather up political and financial information about you to sell it to unknown parties.

Paranoia about such computer intrusions is counterproductive. If you have some personal data like a financial planning system, or a record of your tax returns, or medical information you really want to keep private, you don't need any of the information in my book. What you should do is buy some used computer from a friend or a garage sale. Remember: PCs with an Intel 486 chip used to be really neat machines just a few years ago; you can probably get one for $50 and it will go on working for years! Install it in the basement as your personal machine. Ignore the ribbing from your technically savvy brother-in-law with all the latest gadgets. Do what the government does with its really secret stuff: make sure it isn't connected to anything.

As for your everyday web machine, don't use it until you have taken some elementary precautions, starting with antivirus software.

Antivirus Software

At this writing, the two most convenient packages for virus detection are Norton's Antivirus 2001 software ($40) and PC-cillin 2000 from TrendMicro ($30 as a download). Also in this category are McAfee's products.

These systems are only effective if they are updated once a week and if you let them scan your disks regularly.

Most users of Windows don't need file and printer sharing, two features that can (and should) be turned off in your Control Panel. Experts such as George Kurtz have determined that making sure such features are disabled removes your greatest vulnerability to penetration.

Home Firewalls

In large corporations, firewalls are separate computers that handle, identify, and control every packet transmitted to and from a data center or an Intranet. As an individual, you do not need such a complex system, but the same functions can be executed within the computer you use to access the Internet. A home firewall is a piece of software that filters network traffic, looking for potential attacks from hackers. In particular, it scans for "Trojan horses," pieces of nasty code that get written onto your system for later exploitation by intrusive programs.

A good firewall should keep a log of detected attacks, along with date, time, and the address of the attacking computers. The magazine *PC World,* in its May 2001 issue, tested a number of such products and selected McAfee, Network ICE (Defender 2.1), Symantec (Norton Personal Firewall 2001), Zero-Knowledge (Freedom 2.0), and Zone Labs (ZoneAlarm Pro) as relevant packages. Their preferred choices after testing were BlackICE Defender and ZoneAlarm Pro (each a $40 download) "based on how easy they were to set up and how effectively they provided information to the user." In my experience, the Norton Firewall package (Symantec's Norton Internet Security 2001) is excellent because it also includes antivirus protection. And I like the feature in Norton that also intercepts and kills most online advertising. What a great service!

Some firewalls are precise to the point of being useless. One product I tested on my PC would alert me every time some program on my computer was trying to access a remote site, or vice versa, but it would only identify these sites by their number (their "IP address") so I had no convenient way of knowing whether the request was legitimate or not. You will find that much information goes in and out of your computer to update some parts of the operating system or the web connection, which does not pose any real threat (unless it is a bogus update with a Trojan Horse!). The challenge is to separate such requests from harmful intrusions.

Just to put things in perspective, cyber threats reported to Carnegie-Mellon numbered less than 500 in 1991, more than 4,000 in 1998, and over 22,000 in 2000. As one of my correspondents remarks, "Culling out serious threats from nuisance intrusions is going to be a huge challenge in the years ahead."

Suspicious E-mail

Today I logged on to my computer after a ten-day vacation and found 249 e-mail messages waiting for me. Some of them are important indeed: they relate to my professional activities or convey news of friends and family. But over half of them are pure junk, even after filtering by the automatic screening system built into my browser. And some may contain hostile "applets" of the kind that may destroy files on my computer, corrupt the operating system, or extract information from my files, sending it to unknown destinations.

As I sit at the keyboard to clean up this mess, my mind is asking some tough questions. Where is the ability to "augment human intellect" we were trying so hard to implement at the dawn of network technology? Where is the vaunted power of man-machine interaction?

Some of the offending messages seem to rise to the level of legitimate inquiry. They even give you a chance to be removed from the list of individuals to whom the company is offering to deliver investment opportunities, low-rate insurance, or attractive mortgages, some of the favorite products of junk mail advertisers. The problem is that by responding to their message with a request to be removed, you are signaling to the junk disseminator that you have read his garbage. This will put you on a list of desirable targets: you actually read that stuff!

There is a finite probability that you may buy something from them the next time they offer you a book on the secrets of fly-fishing, or a rebate on a new Chevy. Don't expect the level of junk mail to go down after you sign off from such a list. The best advice is simply to delete the unread entries without opening them. After all, what are the chances that you would get a legitimate message entitled "I love you" from someone called Hk2ksr7xpv?

Even better, you can place the sender on a list of automatic deletion for the future, in the hope that he doesn't change his identity every time he hits on you. Some services, such as Yahoo!, provide a "bulk mail folder" into which they automatically store suspected junk messages.

About half of the unsolicited mail offers sexual services, dating clubs, Viagra for women, pictures of naked girls, and "special" catalogs. The rest display a bewildering array of offers for real estate, offshore gambling sites, cheap travel, and all the above-mentioned investment "opportunities."

My personal favorite is the guy with the offer for me to "become a kung-fu master in one week."

Beware of the services that tell you that you are a "finalist" in some contest you never entered, or propose attractive software for a surprisingly low price: systems that report on your credit rating, for instance; or programs that will enable you to "uncover your boss's every secret."

These may hide attempts to get you to reveal your credit card number. If you don't know the company making the offer, don't fall for it.

Responding to that last category can get you into real trouble, because the line beyond the hook can reel you into the arms of some unsavory characters.

E-mail Stalking

Many users of the Internet find their names attached to various lists. The offending party often turns out to be a well-meaning friend who is trying to do you a big favor by associating you with his or her own favorite activity. These lists or newsgroups may be very active, and they will fill up your computer with thousands of pages of marginally interesting stuff and unfiltered rumors. How do you stop this flow of well-intended trash from overwhelming your normal workload?

The answer depends on your relationship with the people on the list. If they are colleagues or friends, they will probably understand a sober notice to "please remove me from this list" based on your current activities or lack of time to properly read the messages and respond to them. But there are cases when this does not work. Perhaps the folks in the group really want you among them. Or one person deletes your name from the list but another user responds to an earlier message to which you are still attached, and you find yourself co-opted again. This can lead to really acrimonious exchanges and to outright fury.

I have made it a practice to remove myself from any group that uses the Internet to disseminate rumors or personal attacks against anyone, or to exploit the network as a medium for crusades. In the process I have antagonized some of my friends who feel passionate about such issues. When I find that polite requests for removal are not effective, I use the browser's options to make sure such mail is deleted as it arrives, and is not even stored in the "trash" folder of my e-mail system.

If you use Microsoft Outlook, you can eliminate spam in the following way: select "TOOLS," then "MESSAGE WIZARD," and create a folder for spam, or junk. When you get a frivolous message, right-click it, select "JUNK E-MAIL" and "ADD TO JUNK SENDERS." Any future message from that address will go to the spam file.

The Cookie Jar

Cookies are useful to sites you visit often, such as your bank, online broker, or a common service provider such as Amazon. A cookie is a short sequence of characters that a web server (such as E*Trade or Yahoo!) can store temporarily on your hard disk. The server sends a request to your web browser (typically, Netscape or Internet Explorer) for a cookie to be written after a certain action. For instance, as you make purchases in an online shopping mall, your browser writes the list of items you purchase so that a total can be tabulated from your local "shopping cart."

At the end of every session, the browser erases cookies that have expired. Others are written to a cookie file so they can be recalled at the next session. You can use your text editor to see these bits of data. Some cookies are very helpful. They enable your favorite services (a bank, an online broker) to recognize you on successive sessions and to optimize their user interface based on what you have requested before. However, there is a less savory way for abusers to use cookies. They represent a potential way to find out what you do on the web, how long you stay on each site, and what you do there. For that reason, many groups have requested that cookies be restricted in time and in the ability of unauthorized websites to access them.

As this book is written, cookies are given an expiration date and can be recalled only by a site you have authorized. There is still room for abuse, however. Even today, SBC-Pacific Bell lets advertisers write cookies on my machine and I don't know if, how, and when this information is resold.

It is a simple matter to erase cookies. In Internet Explorer you can go directly to the Windows/cookies folder and delete its content. In Netscape, you can type "FIND" and search for cookies.txt, then get rid of it. This will have the effect of forcing you to reenter some user information the next time you interact with your friendly banker, doctor, or pharmacist, but at least your privacy will be protected.

You can also set your browser to give you an opportunity to view the origin of cookies and only authorize them if you like. For instance, in Netscape Navigator you can go from the "EDIT" menu to the "PREFERENCES" section. Click on "ADVANCED" and turn on the feature called "WARN ME BEFORE ACCEPTING A COOKIE." Be prepared to be amazed at the variety and complexity of the data that underlie your every move on the web.

Caches

Your browser (Netscape Navigator, or Microsoft's Internet Explorer) stores an image of all the pages you visit on the web in order to speed up interactive sessions. When you return to a previously visited site, it will simply recover that page from your hard drive instead of getting a new version from its remote server, unless you specifically instruct it to "RELOAD" or "REFRESH" the display.

Since these caches stay inside your PC's memory beyond the end of the session, anyone with access to your computer can look up this file of caches and find out where you have been. For instance, your service provider could find out you have been visiting the Amazon site or an online auction competitor at eBay. There is no reason for you to give out such information on your behavior as a consumer.

You can erase the file of caches in Internet Explorer by selecting "TOOLS," then "INTERNET OPTIONS" where you pick the "GENERAL" tab. Under "TEMPORARY INTERNET FILES" section, click on "DELETE FILES."

With Netscape you would select "EDIT" and "PREFERENCES." Under "CATEGORY" double-click on "ADVANCED" and select "CACHE." Continue with "CLEAR MEMORY CACHE" and "CLEAR DISK CACHE."

History

Your browser also keeps a history file. This is a log that identifies each website you visit. There is no reason to leave this file open. To purge it under Internet Explorer, select "TOOLS" and "INTERNET OPTIONS" as you did for caches. In the "GENERAL" tab, click the "CLEAR HISTORY" button. The same thing can be done under Netscape through "EDIT" and "PREFERENCES" by selecting "NAVIGATOR" in the "CATEGORY" window, followed by "CLEAR HISTORY."

Other Countermeasures

It is not a good idea to conduct confidential business over the Internet as it exists today, but if you must do it, you should use encryption. Any claim that such software is "impossible to crack," however, should be taken with a grain of salt. If you were an intelligence agency with the capability to decipher common encryption software, why would you ever tell the world about it?

What you can accomplish as a first step is file encryption that is good enough to discourage casual hackers, yet simple enough so you can recover the original text if you have to. This can be done with a technique called, appropriately enough, "pretty good privacy," or PGP, and it is offered in various forms by such companies as CenturionSoft, Network Associates, Panda Software, and CyPost. I have not used any of these, but Stan Miastkowski writing in *PC World* has recommended Steganos Security Suite from CenturionSoft ($60).

As a second step, various experts recommend that you experiment with private e-mail systems such as Sigaba Secure and HushMail. Some services, such as Yahoo!, offer you the convenient ability to encrypt your mail through their secure system.

For everyday secure e-mail with your friends, several services are available. Major portals like Yahoo! already offer a way to send encrypted mail. One convenient service, at www.Hushmail.com, has a very effective mechanism for protecting mail exchanges.

It is not a good idea to place too much trust on e-commerce sites that claim to be secure. A study carried out by N-cipher, a security company, examined 137,000 websites claiming to offer strong encryption as a way of protecting their commercial users. The report[71] found that 19 percent of them could be broken fairly easily. The proportion was 15 percent in the U.S., 19 percent in Britain, and 41 percent in

France. Any key less than 900 bits is regarded as too short to be secure today. The scary thing is that among the easily broken sites were Barclay's Capital, Royal Bank of Scotland, the U.S. First National Bank and Trust, and Royal Bank of Canada. Presumably, they have changed their security software since the report was published.

Web Bugs and Spyware

Ever since cookies have been made safer and more private, advertisers have been looking for new techniques to find out what you do on the web. Some methods are ingenious. For example, you have to admire the company that offered free software to "get rid of advertisements," which has the added advantage of radically speeding up your use of the network. Thousands of Internet users downloaded the program, which would reroute any request they made through its filtering ("proxy") server. What they didn't tell you was that their server logged such requests while eliminating the offensive ads. The company then sold their profile to the same advertisers whose ads had been screened out!

Many innovative companies have found another way of tracking your every move on the Internet through the technique of "web bugs," an image file embedded in a web page, and designed to send back information to the service you are using. web bugs are invisible. They can operate anywhere in your file system, gathering up your stored passwords or looking up your financial data. A security company named Intelytics has found that 75 percent of the major e-business sites were using the technique.[72] Several new laws have been proposed before Congress to rein in these abuses, but this particular horse has already left the barn.

Services that work on behalf of advertisers, such as DoubleClick, are said to be making extensive use of web bugs. You can inhibit this very sophisticated intrusion by turning off "HTML DISPLAY" from your browser, but there is no convenient way to completely block the practice at this point. You can inhibit DoubleClick by opting out of their program monitoring your ad clicks, by going to their site at:

Doubleckick.net/company_info/about_doubleclick/privacy/
privacy2.htm

You can also download Steve Gibson's OPTOUT program, which detects and expunges spyware.

12

To Create a System

A long technical career is a wonderful thing. You get to witness developments you never would have dreamed possible. Thus I had the pleasure, in the late 1990s, of attending a presentation about the Internet at the occasion of a formal conference of the European Union held in Brussels. The speaker was a vice president with IBM, who was explaining the basic process of the Internet and promoting his company's expertise in the new technology.

I savored this presentation in a way that was probably lost on most of the people in the room. I remembered that IBM, like AT&T and most of the world's communications monopolies, had spent the previous quarter-century trying to kill the Arpanet, the Internet, and anything remotely connected with the concept of packet networks, so there was a special sense of pleasure at listening to this new pitch by the giant technology company that was about to teach to a new generation of Eurocrats the finer points of web surfing.

The historical forces behind the rise of the Internet were so powerful that they bent the major industrial corporations and molded their management structures into new shapes.

Other experiences of my career were not so happy. A couple of years ago I arrived at my office in San Francisco to find my secretary very upset. She had found both of our computers in an open condition, with displays lit up as if someone had been using them overnight. As a small venture capital company, we had little in the way of secrets and could hardly attract the attention of sophisticated hackers. Yet someone had clearly obtained a key to my office and perused our files, which we had not bothered to encrypt.

We had been unconcerned with threats to our business, perhaps foolishly. I had thought it was enough to install a couple of security programs, anticipating random amateur attacks through the web. Only, this particular invasion had been a direct, old-fashioned breach of the most elementary security: the key to our office door!

The intruders had not been able to disarm the firewall software, so they left a clear trace of the session duration and the time of their visit. They left with a list of our e-mails, professional files, and a couple of personal records. My computer held a mailing list of all my business contacts, numerous memos about companies we were financing, and a statistical file about parapsychology someone had sent me for review. Hardly the stuff of spy novels.

I have some idea about the perpetrator who hired someone to pay us this nocturnal visit. I was negotiating at the time with two European institutions that had an active interest in high technology. Business practices and rules in Europe are even less careful about individual privacy than in America. As a result, I was led to study a bit more closely how the use of the web was shaping up on the other side of the Atlantic.

The results were not very encouraging.

Superficially, protection of a citizen's rights is absolutely rigid throughout Europe: Every government has well-advertised regulations against database abuses, and the press is full of self-congratulating articles in which America is depicted as a land of many computer intrusions while French and English citizens can rest assured that their medical records are safe, their tax information inviolate, and their romantic e-mail messages unbroken.

It is a very naïve Frenchman or Englishman who believes all that, as we saw in the previous chapter.

In an interview published in the July 21, 2001, edition of *New Scientist*, Caspar Bowden of the Foundation for Information Policy Research in London had this to say about privacy concerns in Britain, where citizens are supposedly protected by the official Regulation of Investigatory Powers Act:

> The powers that are left still provide for massive trawling of communications inside the UK in a way that is totally unprecedented, even in wartime.

If some future British government turned out to be "of an extreme political character," to use governmental language, it could use the powers of the Act to treat every citizen as a potential criminal, under the guise of chasing pedophiles, terrorists, or drug traffickers.

Information is control. French law prohibits companies from prying into the records of citizens, but major advertisers and consumer businesses can routinely hire specialized firms to provide them with databases filled with customer profiles. Some of these records come from questionnaires that people have agreed to fill out in return for some prize or discount; others are compiled as a result of surveys, telephone inquiries, or "special offers" made by marketing firms.

Even if the names of individual customers do not appear in a way that could link these entries together, sophisticated statistical procedures can bring the focus of the analysis to a very fine point indeed: a city block, or an individual building. In the near future, this will all be augmented through the use of web telephony. Under the European GSM telephone system, it is already possible for technicians to locate any cell phone user within a few meters. When the phone becomes a web platform accessing interactive services and making purchases, it will be only a small step for a marketing organization to "suggest" attractive purchases in the vicinity, all in the name of convenience.

At that point, subtle control of an individual's every move, not only virtually but also physically, becomes feasible. Patterns can be detected and predicted. The forecasting of demand for products and services becomes a matter of simple mathematics. As for the monitoring of an individual's moves through the city, from home to office and from the car to the metro or the bus, it can be achieved within the existing infrastructure of the major phone companies—which happen to be the same firms that provide Internet service, and the advertising, and the content, through the ownership of music studios and major publishers.

Availability of webcams equipped with face recognition features, watching streets and malls throughout our major cities, as already practiced in Great Britain, has the potential to create an environment any of the great despots of history from Hitler to Stalin would have relished—and to do so "democratically" under the guise of protecting the safety and comfort of all citizens.

The world where the behavior of large numbers of people can be predicted and controlled exists today. It is not a science-fiction fantasy or an approximation of some Orwellian prophecy. It is both more complex and more subtle than any utopian scenario.

Nobody knows what form the Internet ultimately will take. The future could defeat all the power-grabbing schemes of media companies just as easily as it could play havoc with the projections of idealistic advocates of free information. The inventors of new technologies are of no help in this regard. Thomas Edison thought that his phonograph would be used mainly to relay messages from the telegraph office to individual customers. He did not envision its use for recording music. Alexander Graham Bell, on the contrary, thought his telephone would be used primarily by people in their living rooms, eager to listen to remote concerts.

Something huge is evolving through the multiplicity of interconnecting networks that are being erected all over the planet, and no one has a general map of the new organism. When I asked Paul Baran how he imagined the future of the network he helped create, he reminded me that we had not even reached the point of making broadband services available to consumers. "I've become something of a privacy nut," he told me as we were finishing our Japanese lunch in Menlo Park.

"The real future is in video. If you can give people an upstream path to the net at 4 megabits per second, everybody can have his own TV channel. That's when behavior really changes, lifestyles become different. Five years ago it was too early to talk about video on demand, but now it's becoming a reality. Just look at companies like TiVo and others, that let you take control of your TV set, retrieve shows at will, and skip over the ads. It's still primitive because it has moving parts, but the next generation will be a virtual set. Broadcast is dead: only 15 percent of the U.S. population now get their TV from the air. Cable will have to go digital, as satellite TV has already done. The only technical limitation we still have to conquer is that upstream capacity."

When that prediction becomes fact, the arrow of information will really be reversed. The world will be a very different place when everyone can run his or her own version of CNN, with information streams that will overwhelm officially sanctioned media. The current reality of peer-to-peer computing, instant messaging, free distribution of music and art, and easy sharing of file systems across the world only represents a first approximation of the intellectual universe we can build.

The ultimate control, of course, rests with us. Unlike the old, inaccessible structures of communications that were housed in fortified buildings and relied on billion-dollar electronic facilities and guarded towers at the tops of mountains, the structures of the Internet are virtual constructs we can learn, manipulate, and bend to our designs from the simple, everyday environment of homes and offices. No matter what digital information has been accumulated in databases, no matter what inferences have been made about us by businesses or government organizations in pursuit of their own objectives, the next move is always ours.

As a first step, this book gives a series of suggested measures that will make it more difficult for marketing hacks and snooping data-miners to grab information about you. Beyond these simple measures, the information world you want to inhabit is yours to design.

Originality, unpredictability, and the apparent irrationality of human emotions and desires will become more precious than ever in the world of the web. For that reason, I am alarmed but not over-whelmed by the prospects of interaction with the powerful structures of the future Mesh. If anything, the ubiquitous presence of the net-work will encourage unique thoughts.

The betrayal of the Internet triggers a new wave of concern for freedom, for what it means to be human in a postindustrial society. It exposes the horrible temptations of letting others control our minds. It educates us to the power of our own talents, of that spark of human genius that has never been enslaved by any regime or system.

It is that genius that created the web against all the forecasts, against major industrial interests, against the grand schemes of governments. That same genius can foster network technology to return control to the users themselves, to release their full potential, to link together the thoughts and the aspirations of human beings all over the Earth. But the burden is on us, as individual consumers of these new services, to be sharply selective in the tools we use and the relationships we form in this new medium.

William Blake put it best when he wrote:[73]

> *I must create a System,*
> *Or be enslaved by another man's!*

Endnotes

1. In contrast, for an accurate reconstruction of this technology's development, see "A Partial History of the Internet" by Anthony R. Curtis of the Union Institute, 1997; www.tui.edu.
2. Thomas Belden and Marva Belden. *The Lengthening Shadow*. Boston: Little Brown, 1962.
3. Stafford Beer. "Questions of Quest" in *Cybernetics, a Source Book,* edited by Tobert Trappi.
4. Jean-Pierre Dupuy. *The Mechanization of the Mind*. Princeton: Princeton University Press, 2000.
5. Igor Alexander. Book review in *New Scientist*, Mar. 17, 2001, p. 56.
6. Jerry M. Rosenberg. *The Computer Prophets*. New York: Macmillan, 1969.
7. David W. Gardner. "Will the Inventor of the First Digital Computer Please Stand Up?" *Datamation*, Feb. 1974, pp. 84–90.
8. Christopher Evans. *The Mighty Micro*. London: Victor Gollancz Ltd., 1979 (published in the U.S. as *The Micro Millennium,* New York: Washington Square Press/Pocketbook, 1981).
9. Anthony Cave Brown. *A Bodyguard of Lies*. New York: Bantam, 1976.
10. News item reported in *New Scientist*, Oct. 28, 2000, p. 19.
11. Andrew Hodges. *Alan Turing, the Enigma*. New York: Simon & Schuster, 1983, p. 363.
12. Jerry Mander. *In the Absence of the Sacred*. San Francisco: Sierra Club Books, 1992.
13. Editorial in *New Scientist,* Oct. 13, 2001, p. 3.
14. E. M. Forster. "The Machine Stops" in *The Eternal Moment and Other Stories*. New York: Harcourt, 1929.
15. Robert Horwitz. "Tuning In to WARC," *Co-Evolution Quarterly*, summer 1979.
16. Ray Brosseau and Ralph K. Andrist. *Looking Forward: Life in the Twentieth Century as Predicted in the Pages of American Magazines from 1895 to 1905*. New York: American Heritage Press, 1970.
17. Ibid.

18. Bernstein et al. *Silicon Valley: Paradise or Paradox? The Impact of High-Technology Industry on Santa Clara County.* Mountain View: Pacific Studies Center, 1980.
19. Katie Hafner. "A Paternity Dispute Divides Net Pioneers," *New York Times,* Nov. 8, 2001.
20. Paul Baran. *The Beginnings of Packet Switching—The Underlying Concepts.* Franklin Institute and Drexel University Seminar on the Evolution of Packet Switching and the Internet, Philadelphia, Apr. 25, 2001.
21. Steward Brand. "Wired Legends: Founding Father," *Wired,* Mar. 2001.
22. Katie Hafner and Matthew Lyon. *Where Wizards Stay Up Late.* New York: Simon & Schuster, 1996.
23. Steward Brand. "Wired Legends: Founding Father," *op. cit.*
24. D. Engelbart and staff. *Advanced Intellect-Augmentation Techniques* (Report to NASA under Contract NAS1-7897). Menlo Park: SRI, July 1970.
25. D. Engelbart and staff. *Computer-Augmented Management—System Research and Development of Augmentation Facility* (Report to ARPA under Order no. 0967, Contract F30602-68-C-0286). Menlo Park: SRI, Apr. 1970.
26. *Scenarios for Using the Arpanet, at the International Conference on Computer Communication,* Washington, D.C., Oct. 24–26, 1972). Menlo Park: SRI, NIC publication 11863 (with an introduction by Robert Metcalfe).
27. Jacques F. Vallee. "A Facility to Interrogate Arpanet Resources," SRI-ARC Online Journal, entry 11136, ARPANET, July 1972.
28. Robert W. Chambers. *The King in Yellow.* Freeport, N.Y.: Books for Libraries Press, 1969.
29. *Travel-Communications Relationships.* Transcript of a computer conference, edited by the Institute for the Future. Montreal: Bell Canada Business Planning Group, July 1974.
30. Jacques F. Vallee, G. Askevold, and T. Wilson. *Computer Conferencing in the Geosciences* (Report to the U.S. Geological Survey). Menlo Park: Institute for the Future, Sept. 1977.
31. *International Computer Conference on Psychic Phenomena* (Unpublished transcript). Menlo Park, May–June 1975.
32. The InfoMedia staff at the time included Richard Miller, Robert Verhey, Ruth E. Smith, Sal Suniga, Jennifer Lear, Lindsey McLorg, Carol Sciutto, Jean Dawson, Cos Nista.
33. Florence Amalou. *Le Livre Noir de la Pub: Quand la Communication Va Trop Loin.* Paris: Stock, 2001.
34. Warren Wagar. *Building the City of Man.* San Francisco: W. H. Freeman & Co: San Francisco, 1971.
35. Jacques F. Vallee. *The Network Revolution—Confessions of a Computer Scientist.* Berkeley: And/Or, and London: Penguin, 1982.
36. Lawrence Lessig. "The Internet's Undoing," *Financial Times,* Nov. 29, 2001.
37. Author's interview with Michel Serres, Paris, June 22, 2001.
38. Pam Sword. "Packet Switching Inventor Aims to Make Planet a Better Place," *Halifax Chronicle-Herald,* June 22, 1991.
39. Wendy Grossman. "First Person," *New Scientist,* Sept 1, 2001, p. 47.

40. Kathy Watson. *New Scientist,* Mar. 30, 2002, p. 88.
41. Brad Grimes. *PCWorld,* May 2001 p. 98.
42. Michel Charasse. *Le Figaro,* Nov. 18–19, 2000, p. 16.
43. *Red Herring.* Jan. 16, 2001, p. 51.
44. Jeffrey T. Richelson. *The Wizards of Langley.* Cambridge, Mass.: Westview Books, 2001.
45. Benny Evangelista. *San Francisco Chronicle,* July 30, 2001.
46. Paul Krugman. *New York Times,* Oct. 4, 2000, p. A35.
47. Annick Jesdanum. "Web's Inventor Trying to Keep Things Simple," Associated Press and *San Francisco Chronicle,* Dec. 26, 2000.
48. "Out of the Shadows," *New Scientist,* Aug. 26, 2000.
49. Katie Hafner, "A Paternity Dispute Divides Net Pioneers," *op. cit.*
50. Author's interview with Paul Baran, Menlo Park, Nov. 13, 2001.
51. D. C. Denison. "And the Next Big Thing is . . . ," *Boston Globe,* Oct. 8, 2001.
52. Larry Roberts. "Internet Chronology," Mar. 22, 1997, updated Oct. 24, 1999. Quoted by permission from Dr. Roberts's website.
53. Robert Salladay: "Consumer Privacy Bill Could Be Weakened," *San Francisco Chronicle,* Aug. 3, 2001. p. 1.
54. *New Scientist,* July 29, 2000, p. 4.
55. *Boston Sunday Globe,* Aug. 26, 2001.
56. *Financial Times,* Sept. 1, 2001, p. 7.
57. *The Sciences Magazine,* New York Academy of Sciences, Nov–Dec. 2000, p. 7.
58. Reported in *San Francisco Chronicle,* July 2, 2000, p. A-15.
59. Quoted in *International Herald Tribune,* Dec. 6, 2000.
60. Christopher Dore in *Le Figaro Economie,* Nov. 22, 2001.
61. Verne Kopytoff in *San Francisco Chronicle,* Aug. 30, 2001, p. 1.
62. "Rebels in Black Robes Recoil at Surveillance of Computers," *New York Times,* Aug. 8, 2001.
63. Jeffrey Rosen, *The Unwanted Gaze.* New York: Vintage, 2001.
64. David Burnham. *The Rise of the Computer State.* New York: Vintage, 1984.
65. Dan Hester in *TechWeek,* Oct. 2, 2000.
66. Lisa Bowman. ZD Net news, July 20, 2001.
67. Jacques Ellul. *The Technological Society.* New York: Vintage, 1964.
68. "Tangled Webs," *The Economist,* May 25, 2002, p. 67.
69. Senate Hearings, 92nd Congress, 1st Session, 1971.
70. *New Scientist,* April 27, 2002, p. 7.
71. *Les Echos.net,* April 30, 2001, p. 1.
72. Max Smatannikov. "Beyond Carnivore: FBI Eyes Packet Taps," *Interactive Week,* Oct. 18, 2001.
73. Blake, William. *Jerusalem,* f10.20.

Index of Individuals Cited

Aiken, Howard, 12
Amara, Roy, 48, 80, 83, 89, 95
Amarel, Saul, 86
Ames, Aldrich, 161
Andreessen, Marc, 108
Atanasoff, John Vincent, 10
Atanasoff-Berry Computer, 11
Bach, Richard, 89
Baran, Paul,
    AT&T refusal, 98
    bio, 53-57, 142
    dream of, 119
    "fishnet," 91
    group communications, 80–82
    "groupware," 92
Beer, Stafford, 8, 193
Beren, Peter, v, xv
Berman, Jerry, 172
Berners-Lee, Tim, x, 107, 108, 140
Berry, Clifford, 11
Bhushan, Abhay, 76
Bina, Eric, 108
Bitzer, Donald L., 31
Bowden, Caspar, 152, 172, 188
Breck, Renwick, 102
Brown, Anthony Cave, 10, 13, 193
Burnham, David, 158, 163, 195
Bush, Vannevar, 47, 107, 134–135
Carey, Professor James, 27
Cerf, Vint, 52, 77, 83, 90–92
Charasse, Michel, 133
Clinton, Bill, 141
Cohen, Danny, 91
Crane, Hew, 112
Davies, Donald, 53–54, 142
Davis, Ruth, 88
Doctorow, Cory, 172–173
Doerr, John, 107, 109
Doriot, General, 93
Dupuy, Jean-Pierre, 8, 193
Eckert, Presper, 10
Ellis, James, 152
Ellul, Jacques, 160, 196
Engelbart, Douglas, 46

Erhard, Werner, 64
Evangelista, Benny, 136, 195
Evans, Christopher, 12, 193
Feigenbaum, Ed, 86
Feinler, Elizabeth "Jake," 63, 69, 77, 106
Filo, David, 109
Fogel, Jeremy, 137
Forest, Lee de, 45
Forster, E. M., 37, 193
Gates, Bill, 117
Gore, Al, x, 141
Grimes, Brad, 132, 195
Grossman, Wendy, 125–126, 195
Hafner, Katie, 53, 56, 142, 194–195
Hanssen, Robert Philip, 161
Herzfeld, Charles, 56
Hester, Dan, 159, 196
Hewlett, Bill, 45
Hibbard, Angus, 39
Hynek, J. Allen, 23
Irby, Charles, 69
Jackson, James, 157
Jichuan, Wu, 155
Jinbo, Wang, 156
Jobs, Steve, xv, 117
Johansen, Robert, ii, v, 85, 98
Kahn, Bob, v, 52, 73, 77, 83, 90–91, 106
Keen, Harold, 16
Kertecz, Francois, 34
Kleiner & Perkins, 107
Kleinrock, Leonard, 53, 73
Knox, Alfred Dilwyn, 14
Koch, Hugo, 13
Kozinski, Alex, 157
Krugman, Paul, 137, 195
Larson, Earl, 10
Lehtman, Harvey, 69
Lessig, Lawrence, 117, 195
Licklider, J. C. R., 52, 73, 83, 87, 143
Lipinski, Hubert, 81, 85, 87
Lukasik, Steve, 84
Lynch, Dan, v, 108
Lyon, Matthew, 14, 56
Machapetris, Paul, 106

Mander, Jerry, 28, 41, 193
Mauchly, John, 10–12
McCain, John, 131
McCarthy, John, 86
McCulloch, Warren, 8
McIver, Peter J., 17
McLindon, Connie, v, 84, 106
Menzies, Sir Steward, 13, 16–17
Metcalfe, Bob, 76, 117, 143
Miller, Rich, v, 81, 87, 102
Neumann, John Von, 7, 11, 22, 24
Norris, Bill, 97
Packard, David, 45
Partridge, Craig, 106
Pinkerton, Brian, v, 109, 157, 174
Postel, Jon, 91, 106
Pouzin, Louis, 105
Rand, Sperry, 10
Rech, Paul, 74, 85
Rheingold, Howard, 103
Roberts, Larry, 57, 143–144
    ARPA order, 56
    internet father, 52
    message system, 79
    photo, 77
    Telenet president, 84
Rohde, Gregory, 156
Rosen, Jeffrey, 158, 195
Rosenberg, Jerry, 10
Rosenberg, John Paul, 64
Rosnay, Joël de, 36
Schroeder, Mary, 158
Speier, Jackie, 149
Spielberg, Steven, 23
Swann, Ingo, 89
Swanson, Rowena, 48
Taylor, Bob, 52, 54, 57, 73, 142
Teller, Edward, 11, 162
Terman, Frederick, 45
Tomlinson, Ray, 79
Turing, Alan, 10, 14, 17–18, 139, 193
Uncapher, Keith, 86
Wagar, Warren, 111, 195
Watson, Dick, 85
Wiener, Norbert, 8
Williams, Ray, 93
Yang, Jerry, 109
Zuse, Konrad, 12, 16

Index of Topics

@ sign, 79
AbiliTec, 133
Acxiom, 133
Advanced Micro Devices, 46
advertising,
    abuses, 131
    contextual, 136–137
Air Force Office of Scientific Research, 48
Altavista, 110, 135
Amazon.com, 132, 168
Amdahl, 93
America Online, 102, 109
American cybernetics, xiii
antivirus software, 180
AOL,
    See also instant messaging icon
    electronic media rulers, 166
    Netscape merger, 165
    research, 127
AOL Time Warner, 125, 166
Apple, 10, 18, 92–93, 115, 117
ARC (Augmentation Research Center),
    See also Arpanet
    financing of, 58
    initialization of, 46
    "pointer" development, 49
    profoundness of, 50
    programmer's commune, 59
    secretaries of, 63
    selling of, 68
    system mastery, 80
ARPA (Advanced Research Projects Agency), 52–54
    See also Arpanet
    community contact, 67
    computer budget, 63
    EST funding, 68
    government funding, 105
    network sites, 69
    project monitors, 74
    research proposals to, 82
Arpanet,
    birth of, 57
    concept, 57
    control transfer, 91

    death of, 106
    defined by, 73
    denied access to, 104
    first demonstration of, 74
    genesis of, 55
    history of, 75, 90
    how it works, 57
    internet ancestor, 53, 73, 104
    military, and the, 142
    organic growth, 77
    photo, 77
    point of, 120
arrow of information, 109–110, 118–119, 130, 190
artificial,
    intelligence, 27, 35, 40, 86–87
    language, 11
Artificial Intelligence Lab, 26
AT&T, 55-56, 98, 115, 117, 120, 166, 187
Atanasoff-Berry Computer, 11
Atomic Energy Commission, 34
"Augmentation Research Center," *See* ARC
Automatic Calculating Engine, 18
automatic digital machines, 10
Bell Labs, 56, 140
Berners-Lee's protocols, 108
"Bernoulli equation," xvii
Bertelsmann, 125, 150, 166, 172
binary number system, 12, 16
Bodyguard of Lies, A, 13
British Tabulating Machine Company, 16
broadband DSL, 120
browser software, xiv
"browsing," 78
Building the City of Man, 195
bulletin boards, 40, 92, 156, 158, 174
Bureau of Justice Statistics, 29
Burroughs, 57
business,
    online, 132
    practices, 167, 188
buyer profiles, 132
caches, 145, 169, 184–185
Canal Plus, 133
Carnivore, 157, 159–160, 172

cellular phones, 4
censorship, 137, 156, 173
Center for Democracy and Technology, 172
Centre d'Études et Recherches Nucléaires (CERN), 104, 107, 140
CERN protocols, 108
Chess, 76, 87
Chicago Telephone Company, 39
China, 114, 132, 155–157
CIA, 127, 155, 161
"circuit switching," 53
Cisco, 7, 107
"cloning of cultures," 41
Close Encounters of the Third Kind, 23
CODIAK, 113
Colossus engines, 18
commercial channels, xii
communication,
    future, 36
    media, 4
    structures, 55
    systems, 118
Compuserve, 92
"computable numbers," 14
computer(s),
    conferencing, 82
    early, 12
    intrusions, 179, 188
    technology, 21
conferencing, 79
connectivity, 56, 173, 175
control, 8–9
Control Data, 74, 95, 97
"cookie," 128–129, 183–184
cookie jar, 183
Corporation for National Research Initiatives (CNRI), 106
creating languages, 11
creativity, 37, 56, 115
cryptology, 13
CSNET, 104
culture of information, 141
customer profiling, xii
"cybernetics," 8, 130, 193
Cyclades, 106
dark matter, 166–167
DARPA, 91
data rate, 142
database,
    abuses, 188
    building, 134
    designing, 69
    generalized, 77
    hacking into, 155
    information, 94
    simple, 38
data-mining, 148
DCS1000, 160
Defense Communications Agency, 56, 91
"Delphi technique," 81
Digital Equipment Corporation, 57, 93
digital networking, 5

digital society, 26–42
Disney, 40, 45, 127, 132, 166
distributed adaptive message block switching, 53
DNS scheme, 106
Domain Name System (DNS), 106
dotcom, 106, 120, 131, 147
    companies, 38, 130
    crash, 130
DSL (digital subscriber line), 144
"Dynamic Knowledge Repository," 113
eBay, 127, 132, 179, 184
"Echelon," 160
Electric Power Research Institute, 102
electronic,
    commerce, 5, 69, 105, 130
    education, 30
    mail, *See* e-mail
Electronic Frontier Foundation, 172–173
Electronic Privacy Information Center, 172
e-mail,
    network, 79
    private systems, 185
    stalking, 182
    traffic, 159
Eniac, 10–11, 13, 22
Enigma, 13–14, 16, 18
"ENQUIRE," 107
entertainment companies, 40, 117
erosion of privacy, 163
Esalen, 64
EST, 64–68
    experiment, 68
Ethernet cables, 141
    European GSM telephone system, 189
Fairchild, 46
Fall Joint Computer Conference, 48
FBI, 127, 132, 139, 159–161, 172, 178
Federal Telegraph Company, 45
Federal Trade Commission, 135
first electronic,
    brain, 13
    computer, 9
floating-point arithmetic, 12, 16
Foundation for Information Policy Research, 172, 188
free information sharing, xii
freedom of,
    access, 168
    information, xvii
future mesh, 139–150
Future of Ideas, The, 117
General Electric, 96
geodesic organization, 118
gigabit modems, 7
"Global Grid Forum," 145
Global International Liberty Campaign, 153
Global Internet Policy Initiative, 172
Gnutella, 126
Government Code and Cipher School (GC&CS), 14
"grapevines," 119
graphic interfaces, xiv
Grid, the, 144
groupware, 79

hacking, 128, 150, 155, 160
hardware, 5
Hewlett-Packard, 45
history of scientific discoveries, xvii
home firewalls, 180
Honeywell, 10
"hot potato" idea, 55
Hotbot, 135
"HTML," 107–108, 186
"HTTP," 107–108, 120
human,
 mind, 19, 25, 37, 110
 side of the technology, 6
 touch, 40
hyperlinks, 107, 136
hypertext, 107
IBM 360, 93
IBM 650, 9, 19
ICANN (Internet Corporation for Assigned Names and Numbers), 156
In the Absence of the Sacred, 28
InfoMedia, 92–95
Information Processing Techniques, 54, 56
information science students, 10
infrastructure, xii
In-Q-Tel, 157
instant messaging, 4–5, 82, 94, 135, 158, 190
"instant messaging" icon, 135
Institute for the Future, 81
Intel, 46, 117, 167, 179
intellect enhancement, 60
International Broadcasting Bureau, 156
Internet, 92
 Explorer, 108, 110, 135, 183–185
 messaging, 160
 revolution, xii
 service provider (ISP), 127
 Society, 108
INTERNIC, 69
"invisible colleges," 43, 119
Iowa State University, 11
IP,
 address, 177, 181
 dial-tone, 174
Jet Propulsion Lab, 23
"Jotto," 76
KaZaa, 136
Kilobit modems, 7
Knowledge Revolution, 63
Knowledge workers, 59–61, 63
law enforcement agencies, 29, 133, 152, 158
laws of automata, 10
lawyers, 133
"Life," 76
Linux, 145, 175
Lipinski, Hubert, 81, 85, 87
literature of computing, 6
Looksmart, 135
Lycos, 135
machine translation, 33–34
Macintosh, 115
Madam (Manchester Automatic Digital Machine), 18

Mark I, 12, 20
MCI, 104
MCI Mail, 92
Mechanization of the Mind, The, 8
megabit devices, 7
"Memex," 47, 134–135
"message block," 53, 142
META-NICs, 113
Microsoft, 35
 Outlook, 183
 Passport system, 135
 Word, 35
Mighty Micro, The, 12, 193
military bureaucracy, 91
Minitel, 54, 105–106
monopolies, 41, 105, 166, 172, 175, 187
Mosaic, 108, 130, 141
MP3.com, 125
Multics, 78
multimedia games, 7
municipal networks, 173
Napster, 125–126
NASA, 62, 100–102, 108
National Bureau of Standards, 88, 95
National Center for Supercomputing Applications, 108
National Crime Information Center (NCIC), 29
National Science Foundation, 81–82, 88, 104, 130
National Security Agency (NSA), 160
national security issues, xi
NATO, 154
Naval Ordnance Laboratory, 11
Navigator, 108–109, 184–185
N-cipher, 185
net-as-broadcasting scenario, 126
Netscape, 108
network culture, 82
Network Information Center, 69, 74, 76–78
Network Measurement Center, 76
network processes, 91
Network Revolution, The, 117, 195
network technology, 4, 87, 100, 129, 174, 181, 191
networking, 5
networks, 4
neutrality, 56, 169
New Age consciousness, 64
New York Times, The, 53, 137, 142, 158
News Corp, 166
newsgroups, 40, 129, 182
NICs (Networked Ics), 113
NLS system, 50, 60, 69, 74
notepad, 100, 102
NSA intelligence, 140
NSFNET, 105–106
Oak Ridge National Laboratory, 34
office automation, 60–61, 63, 97
Office of Advanced Analytical Tools, 134
online file-sharing program, 136
"Open Systems Interconnection," 105
packet switching, 52–57, 80, 142
PAFree.net, 174
Palm Pilots, 4
paranoia, 29, 76, 127, 179

Paris Observatory, 9, 19, 21, 25
"peer-to-peer" concept, 145
Pentagon, 48
personal,
    countermeasures, 177, 179, 181, 183, 185
    internet servers, 173–174
    networks, 5
Pew Internet and American Life Project, 130
PGP, 185
Planet, 88–89
Planet System, The, 87, 92, 95
Plato systems group, 97
Polaris Venture Partners, 143
"positive thinking," 64
Prestel system, 106
primitive teaching machines, 35
Privacy Foundation, The, 172
privacy,
    invasion, xi
    investigators, 133
productivity, 61, 115, 165, 167
protection of privacy, 30, 157, 168
RAND corporation, 33, 53
RAS (Redundant Acronym Syndrome), 50
Real Networks, 125
RealAudio links, 137
RealNetworks, 135
"record carrier," 94
Red Herring, 133
Research Council of Canada, xiv
research facilities, 104, 108
Rise of the Computer State, The, 158, 163
Safeweb, 156–157
satellites, 9, 21–23, 37, 100, 146
SBC Internet Services, 128
Scientology, 64
search engine bias, 135
security, xi
Senate Commerce Committee, 131
Seven Brothers of electronic media, 166
Shannon's law, 140
Shockley Transistor Corporation, 46
"Silicon Gulch," 51
Silicon Valley, 44
software, 4–5
    "locator," 78
    start-ups, 95
"Solid State Society," 4
Sony, 125, 150, 166, 172
special interface machines, 58
spyware, 186
SRI International, 46, 50–51
SRI-ARC, 47
Stanford,
    AI Lab, 27
    Miracle, 45
    Research Institute (SRI), xii
state-sponsored piracy, 153
stored program, 11
structural integrity, 169
supercomputers, 145
symbolic algebraic manipulator, 76
TCP/IP, 91, 104–105

technobabble, 6
technology, 4
TechWeek, 159
Telemail, 92
telephone, 39
    communications, xi
    unlisted numbers, 133
television, 41
    digital, 173
    interactive, 106, 117, 144
    live, 110
    stations, 136
    text display, 49
Tenex, 78
Texas Instruments, 94, 96
Thomson-CSF, 102
time-sharing resources, 105
TOP-text, 136
Toysmart.com, 132
Transmission Control Protocol (TCP), 90
"Triangle Boy," 157
U571, xv, 14
"ultimate information machine," 119
"Unfinished Revolution, The," 112
universal access, 125
"universal resource locators," 107
Unix, 78, 175
Unwanted Gaze, The, 158
user profiles, 148
USGS, 89, 95, 98
Varian Associates, 45
Viacom, 166
video,
    downloading, 120
    streams, 7
videoconferencing, 41, 117–118
    systems, 117
virus, 128, 131, 139, 180
Vivendi, 125, 127, 133, 150, 166, 172
Vivendi Universal, 166
Voice of America, 156
"Water Buffalo" pitfall, 33
"Web, the," xi
web,
    browser, 108, 141, 183
    bugs, 149, 186
    telephony, 120, 189
webcam, 135
Webcrawler, 110
Western Union, 94, 96–97
Where Wizards Stay Up Late, 56
WiFi network, 174
Windows, 5, 81, 135, 180, 184
World Trade Center attack, 139
World Wide Web, 107
Xerox, 63, 74, 76, 88, 115
Xerox Palo Alto Research Center, 76, 88
X-rated websites, 139
Yahoo, 109, 137
Z3, 12, 16
Z4 model, 12
Zen, 64

About the Author

In his role as a general partner with a Silicon Valley-based venture capital group, Dr. Jacques Vallee invests in young enterprises in Europe and the United States. An early participant in the development of the Internet, he has authored twenty books and over fifty articles in fields ranging from astrophysics to database management, communications, and finance. Although he spearheaded his group's investment in a dozen companies now traded on Nasdaq, Jacques is not a financial executive recently converted to the wonders of technology. Educated in France, where he received a master's degree in astrophysics, he specialized in computer science in the United States, obtained his doctorate in artificial intelligence at Northwestern University, and led the team that developed pioneering network software on Arpanet, the ancestor of Internet.

A passionate observer of Silicon Valley, he founded an information technology company there before becoming a venture capitalist. He lives in San Francisco with his wife, Janine. They have two children.

Jacques Vallee can be reached at: documatica@aol.com.

Russell Targ Editions

The Heart of the Internet by Jacques Vallee

UFOs and the National Security State by Richard Dolan

Studies in Consciousness/
Russell Targ Editions

Mental Radio by Upton Sinclair

An Experiment with Time by J. W. Dunne

Human Personality and Its Survival of Bodily Death
by F. W. H. Myers

Mind to Mind by René Warcollier

Experiments in Mental Suggestion by L. L. Vasiliev

Mind at Large
edited by Charles T. Tart, Harold E. Puthoff, and Russell Targ

Dream Telepathy
by Montague Ullman, M.D., and Stanley Krippner, Ph.D.
with Alan Vaughan

Some of the twentieth century's best texts on the scientific study of consciousness are out of print, hard to find, and unknown to most readers; yet they are still of great importance, their insights into human consciousness and its dynamics still valuable and vital. Hampton Roads Publishing Company—in partnership with physicist and consciousness research pioneer Russell Targ—is proud to bring some of these texts back into print, introducing classics in the fields of science and consciousness studies to a new generation of readers. Upcoming titles in the Studies in Consciousness series will cover such perennially exciting topics as telepathy, astral projection, the after-death survival of consciousness, psychic abilities, long-distance hypnosis, and more.

Hampton Roads Publishing Company

. . . for the evolving human spirit

Hampton Roads Publishing Company
publishes books on a variety of subjects including
metaphysics, health, visionary fiction, and other related topics.

For a copy of our latest catalog,
call toll-free, 800-766-8009,
or send your name and address to:

Hampton Roads Publishing Company, Inc.
1125 Stoney Ridge Road
Charlottesville, VA 22902
e-mail: hrpc@hrpub.com
www.hrpub.com